HOW
LIFE
WORKS

HOW LIFE WORKS

The Inside Word from a Biochemist

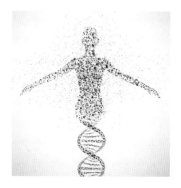

William Elliott and Daphne Elliott

CSIRO

PUBLISHING

National Library of Australia Cataloguing-in-Publication entry

Elliott, Daphne C., author.

How life works : the inside word from a biochemist / Daphne C Elliott and William Elliott.

9781486300471 (paperback)
9781486300488 (epdf)
9781486300495 (epub)

Includes index.

Biochemistry – Popular works.
Life – Origin.

Elliott, William H., author.

572

Published by

CSIRO Publishing
Locked Bag 10
Clayton South VIC 3169
Australia

Telephone: +61 3 9545 8400
Email: publishing.sales@csiro.au
Website: www.publish.csiro.au

Front cover and title page: Abstract model of woman and DNA molecule. Source: Lonely/Shutterstock.com.

Edited by Elaine Cochrane
Cover design by Alicia Freile, Tango Media
Typeset by Thomson Digital
Printed in China by 1010 Printing International Ltd

Original print edition:
The paper this book is printed on is in accordance with the rules of the Forest Stewardship Council®. The FSC® promotes environmentally responsible, socially beneficial and economically viable management of the world's forests.

MIX
Paper from responsible sources
FSC® C016973

Foreword

William Herdman (Bill) Elliott was born in County Durham and graduated in Biochemistry from Cambridge University, where he was elected to a Fellowship at Trinity College. He spent several research years in Cambridge, Harvard and Oxford, and immigrated to Australia in 1957. After eight years at the Australian National University, during which he was elected a Fellow of the Australian Academy of Science, he was appointed Professor of Biochemistry at the University of Adelaide, a position he held with distinction for 23 years; by his leadership he made his Department significant in the world of biochemistry. During that time I was fortunate to be a member of his staff, and all of us worked with him in developing and enjoying the adventures of teaching and research. I had an enduring friendship with Bill after his retirement, and memories of his humanity and good humour remain strong.

Bill's ability as a researcher was equally matched by his talent as a teacher who, enthused with his subject, attracted countless students to biochemistry. He displayed the breadth of his biochemical knowledge and writing skills in his widely successful textbook (*Biochemistry and Molecular Biology*, OUP) co-authored with his wife Dr Daphne Elliott. It is now in its 5th edition, and reviews have noted the style of writing as 'refreshingly different, more personal and more direct that seems to take the reader on a one-to-one tutorial'. Others noted the approachable style, and the skill in organising and explaining complex material in an accessible way.

For many years Bill was eager to write about the biochemical basis of living processes in a way that would appeal to a general readership; a notable challenge. The result is this book, an impressive piece of scientific story-telling that deals with how living cells accomplish their remarkable processes of growth and maintenance; he describes them as 'majestic'. His enthusiastic writing, carefully crafted for understanding, is flavoured with anecdotes including his personal acquaintance with famous scientists. The conversational style, characteristic of his textbook, together with thoughtful figures and extra information in text boxes, adds to the quality of the book. Apart from general appeal, as referred to in the Preface, the style of this book should attract the attention of senior school biology teachers. It is a tribute to Professor Elliott's character that he had the will, again with the aid of his wife, to complete the work when in the last months of his life in 2012.

Professor George E. Rogers; AO DSc FAA
Honorary Research Fellow, Molecular and
Biomedical Science
University of Adelaide

Contents

Preface

This book aims at explaining the fundamentals of life to readers who have had no scientific training. It makes no mention of philosophical or spiritual aspects but is concerned only with the mechanism of life based on experimental science. In this sense, life is a process based on large numbers of chemical reactions, which in essentials are much the same in all life forms, from bacteria to humans and plants. Many people do not know what a chemical reaction is, or indeed what a chemical is. This would appear to preclude them from understanding the process of life at anything but a superficial level.

Behind the detailed chemistry of living things there is the big picture of how the establishment of life had to cope with the laws of the universe. This is what the book aims at explaining. It deals in succession with the major problems that life had to overcome by travelling along the road from 'atoms' to the complexity of fundamental biological processes, and tackling, in turn, the points along this progress where fundamental obstacles have been overcome in arriving at 'life' as we know it.

The solutions to these problems are fascinating and indeed majestic, and can be presented in a way that is accessible to non-scientific members of the public. My aim is to explain the quasi-miraculous process by which inanimate matter becomes living matter. For examples of this, consider the way enzymes have overcome energy barriers, or the way molecules have become the turbines that result in the production of ATP that power so many life processes. The focus on the atomic aspects of these remarkable processes hopefully will remove the mystery while revealing the majesty.

The book opens with an account of the properties of atoms that make life possible, and how they came to be present on Earth. This leads on to the nature of chemical reactions, and why they occur. It progresses through the way living organisms cope with the laws of the universe governing energy, to solving the seemingly impossible way of coping with the barrier to the occurrence of chemical reactions by 'inventing' proteins, arguably one of the most remarkable type of molecules in the universe.

This leads to an explanation of the genetic system and the role of DNA in it.

Finally, the origin of life is discussed.

All of this has a basic relevance to human beings. Most people are interested in the nature of the living process, especially as the knowledge is of ever-increasing importance in medicine, agriculture and indeed in all biological sciences. The book should give an increased understanding of the process of life at a level not usually found in popular literature. It will possibly enable non-scientific decision-makers and the general members of the public to better understand some of the important biological and medical issues that face society.

W.H. Elliott
Adelaide, April 2012

Acknowledgements

When Professor Bill Elliott died in July 2012, he had finished and was happy with the book he had produced over several years. It was always clear to his family that it was going to be important to see that the passion he had put into it was going to get to as wide an audience as possible.

I took advice from Dr Jim Peacock, who was very enthusiastic about the project. I owe him a great debt for his encouragement, and subsequently I obtained a contract with CSIRO Publishing. There were some changes required, mainly the introduction of colour and examples with clinical or health relevance which would resonate with daily life. Several people contributed to the finished book.

I want first to thank Dr Laura Frank, Bill's and my granddaughter, for transforming all Bill's black and white line drawings with colour, and in many other ways supporting me in producing the final manuscript.

Oxford University Press was happy to grant permission to use coloured material from our book,

Biochemistry and Molecular Biology, 4e, Elliott, William H. and Elliott, Daphne C. (2009), specifically the material in the table on page xviii.

My thanks are due to science writer Dr Sarah Keenihan for her sensitive collaboration and excellent input as requested by CSIRO Publishing. The following material has been produced by Dr Keenihan: Boxes 2.2, 2.4, 3.1, 3.2, 3.3, 4.1, 4.2, 5.1, 6.3, 6.4, 7.1, 7.2, 7.3, 7.4, 8.1 and 8.2.

Then there are people who have kindly read parts of the manuscript. Bill consulted widely, especially in writing Chapter 1. Among those whose expertise he sought, and whose ready help was given, were:

Anthony Thomas, Elder Professor of Physics, University of Adelaide
Dr George Grean, Reader in Chemistry, University of Adelaide
The late John Prescott, Professor of Physics, University of Adelaide
Leigh Burgoyne, Professor of Biology, Flinders University.

Finally those who have read parts of the manuscript more recently, during the preparation of the final copy: for their support in this, and throughout the project. I would like to thank:

Lynn Rogers
Dr Jane Elliott
Dr Michael Elliott
Dr David Elliott

Dr Laura Frank
Dr Oliver Frank.

Particularly, my thanks go to Bill's good friend and colleague, Professor George Rogers, School of Molecular and Biomedical Science, University of Adelaide, for the generous Foreword he has written.

A last word of appreciation and thanks goes to the staff at CSIRO Publishing:

Julia Stuthe, Publishing Director
Lauren Webb, Development Editor
Tracey Millen, Editorial Manager

for the support and encouragement they provided in their various roles.

Daphne Elliott, AM, PhD
Honorary Research Fellow,
School of Biological Sciences,
Flinders University
April, 2014

Figures reproduced by permission OUP	Corresponding figures in *How Life Works*
Box 1.1, p. 6	Fig 1.2 H–H covalent bond
Fig 2.3, p. 18	Fig 2.2b Animal cell
Fig 6.2a, p. 89	Fig 3.2 Lock and key model
Fig 6.3b, p. 89	Fig 3.3 Hexokinase combining with its substrate
Fig 6.1, p. 88	Fig 3.4 Energy profiles catalysed and uncatalysed reactions
Fig 4.22, p. 63	Fig 4.1 Space-filling model of haemoglobin
Fig 4.8, p. 53	Fig 4.3 Heat denaturation of a protein
Fig 21.2, p. 318	Fig 5.2a AT–GC base pairing
Fig 31.3a&b, p. 319	Fig 5.2b Double helix structure
Fig 21.5, p. 320	Fig 5.3 Model of B DNA
Fig 21.1, p. 317	Fig 5.4 Hydrogen bonding of Watson–Crick base pairs (modified from original)
Fig 7.2, p. 102	Fig 8.1 Cell membrane structure

1

The fantastic nature of matter

In a book whose main aim, as outlined in the Preface, is to describe the fundamental nature of life, you could reasonably ask why it should start with an account of the atomic nature of matter, meaning anything in the universe that occupies space.

Everything is made of atoms, including all life forms. The existence of the phenomenon we know as life is due to the nature and properties of atoms, so this is a good place to start in understanding what life is. A bonus is that the story of how atoms came to exist on Earth is one of the most exciting in science.

The name 'atom' derives from the Greek word *atomos*, meaning indivisible. The Greek philosopher

Democritus arrived at the concept that everything – all solids, liquids and gases – is made of minute particles that cannot be divided into smaller units. In other words, if you could cut up any matter into smaller and smaller pieces you would ultimately arrive at particles that you could not subdivide any further. Atoms were envisaged more or less as hard solid little balls. Although the concept of atoms has been around for so long, it is only in the last two centuries that they have been scientifically studied, and only in the last two decades that anyone has actually seen them as images using new microscopic techniques. They are indescribably small – the number in a few drops of water is approximately 10 followed by 20 zeros (10^{21}).

Of the 92 naturally different atoms in the universe, only around 25 are found in your body, some in very small amounts. The four elements hydrogen, carbon, nitrogen and oxygen make up about 99% of the total number. The term 'element' is used to refer to a substance that is made from one type of atom and cannot be broken down by chemical means. The atoms in living organisms are the same as the corresponding ones in non-living matter. A carbon atom in your body is the same as a carbon atom in chimney soot or anywhere else in the universe, and the same applies to all atoms. So what then is the nature of life? What makes the difference between non-living and living matter?

We can talk about life as a single process. This may seem surprising in view of the vast variations found in life forms. The French Nobel prizewinner Jacques Monod famously expressed it as 'what holds for *Escherichia coli* is true for an elephant'. He meant that the basic chemistry of the microscopic bacterial cell *E. coli*, which lives in the human gut in countless numbers, is much the same as the chemistry of cells in an elephant. The similarities far outweigh the differences.

Life is basically a chemical process. And this is true for all living organisms. The reason for this is that there was a single origin of life from which all life has developed over approximately 3 billion years.

The salient feature of life is that it reproduces itself. The first form of life must have been a relatively simple self-reproducing collection of atoms on the primeval Earth, and since then life has been handed down from generation to generation. To do this a system of passing on information had to be devised which determines that the offspring resembles the parents. In other words, there had to be a genetic system.

The basic mechanism of the genetic system is the same in an *E. coli* cell as in a human, a whale, a tree, an insect, or any life form you care to nominate. There are superficial differences in detail between the process in *E. coli* and higher forms, but these do not affect the essentials. It seems that life became locked into the method of passing on genetic information to offspring at a very early stage of evolution.

The process of evolution has also superimposed variations on this basis of life that enable organisms to exploit environmental niches. Plants, for example, developed the ability to use sunlight as an energy source; birds adapted to the air and whales to the sea, but it has not changed the basis of life common to all.

There are also fundamental laws of nature that life had to conform to; once again the remarkable solutions to these problems are much the same in all life. We will come to discuss these in subsequent chapters.

The living unit of organisms is the cell

Living cells are, with the inevitable rare exceptions, microscopic structures surrounded by a membrane (Fig. 2.2). In your body there are many trillions of cells. The *E. coli* cell is about 1000 times smaller in volume than animal and plant cells.

Why are cells so small? They have to communicate chemically with the outside world through the membrane. There are mechanisms for transporting chemicals in and out of the cell and for conveying signals into it. For example most chemical messengers such as insulin do not enter cells but deliver a signal to it via receptors in the membrane, which act like aerials. What has this got to do with cell size? There has to be an adequate area of membrane to service the needs of the cell. Minute cells satisfy this requirement because they have a lot of surface in proportion to their volume, but if you increase cell size the volume increases much more rapidly than does its surface area. You quickly arrive at the point at which the membrane area is inadequate to support the needs of the cell. So living cells have to be tiny.

A bacterium like *E. coli* is free living, but to build larger organisms cells are aggregated into more complex organisms. This requires a regulatory system to coordinate the activities of individual cells to the needs of the organism as a whole. In an animal such as a human this becomes very complex indeed. The number of human cells that have to be kept in step with one another vastly outnumbers the whole of the population on Earth, so you can see this is no small problem. Cancer is the result of a cell no longer observing the regulatory rules and going its own way.

So, a living cell is a chemical device. Obviously many will say that life is more than chemistry, but I am talking about the process of life and, whatever views are held, there has to be a mechanism. To most people without any training in chemistry this statement will not be meaningful, but with a simple knowledge of atoms it is possible to understand the nature of life at a profound level.

Biology is dependent on chemistry

Chemistry is essentially a description of how atoms react with one another, and from this life can be seen as the outcome of a large number of chemical reactions occurring in an organised manner in living cells. Here is the basis for the connection between biology and chemistry.

In the case of humans there are somewhere of the order of tens of thousands of different chemical reactions. This involves rough estimates but the number is very large. Behind these chemical reactions there are several fundamental problems that life had to solve to conform to the natural laws of the universe. What these problems are and the way in which they were solved is mainly what this book is about.

The thousands of individual reactions in life are details which do not have to be described for you to understand the principles of the living process. Leaving these aside, we are left with the big picture of life – the major concepts that determine the nature of life and its relationship to the laws of the universe.

The first step is to look at the nature and properties of atoms. A good point to start is to discuss where atoms and the universe came from in the first place.

The Big Bang

Astrophysicists have discovered that 13.7 billion years ago there were no atoms and no universe. At that date the universe came into existence with an indescribable explosion. The physicist Fred Hoyle, in Cambridge, argued tenaciously for an alternative theory that the universe has always existed and always will exist, known as the steady-state universe. He dismissed the explosion theory, describing it, rather derisively, as the 'Big Bang', but evidence rapidly accumulated that the explosion theory is correct. The term Big Bang summarised it so well that it was adopted by its supporters and is now used by everyone.

To return to the Big Bang, what was it that exploded? No one knows, but it was from a source of infinite density and energy content. It contained all the components and energy of the present entire universe. If you reflect that there are hundreds of billions of galaxies in the universe, each with hundreds of billions of stars, it was some condensation! What was there before this 'thing'? Physicists tell us that the question is invalid because there was no 'before', since Einstein's relativity theory shows that time was created at the Big Bang. So was space. The explosion of the Big Bang is not to be thought of as ejecting things into space but as the creation and expansion of space. The space between objects expanded, so that in the course of the explosion they became increasingly separated as the universe expanded.

The Big Bang explosion is believed to have produced a soup of concentrated energy and fundamental particles, the ultimate components of atoms: the temperature initially in the first few seconds was too high for any atomic structures to exist. As the explosion rapidly expanded, it cooled to a point at which subatomic particles were formed and these assembled further into atoms of hydrogen with smaller amounts of helium. Hydrogen is still the predominant atom of the universe.

What are atoms made of?

Although in classical terms atoms cannot be further subdivided, physicists using atom-smashing machines have released smaller subatomic components from them. There are three types of these components, called protons, neutrons and electrons. All atoms are made of them. Hydrogen has one proton and one electron; uranium, the heaviest atom found in nature, has 92 protons and 92 electrons. As you go down the list of atoms from hydrogen to uranium each element has one more proton than the one above it in the list, with the number of electrons always being the same as that of protons. Thus helium, the second element in the list, has two of each. The next one, lithium, has three of each, and so on. Elements that you will be familiar with are carbon, with six protons and electrons, phosphorus, with 15 of each, sulfur with 16, and iron with 26. The number of protons in an element, known as the atomic number, defines what the element is. An atom with six protons is carbon, and so on. We shall talk about neutrons later.

A proton has one positive electrical charge and an electron has one negative charge of the same size, so atoms are electrically neutral. All my professional life I have been familiar with electrical charges in biochemistry, where they are very important, but when I sat down to explain what they are I realised that I didn't know. Most of us take them for granted. A physicist revealed to me that nobody knows what they are. They are unexplained fundamental properties of nature. It is the way nature is. We identify charges by the way they behave in an electric field. Positive charges repel one another but are attracted to the negative pole, and negative charges repel one another and are attracted to the positive, but which is positive and which is negative is an adopted convention. So, in summary, we know how they behave and their importance in chemistry, but not what they are.

Neutrons, as their name suggests, have no electrical charge, but they have almost the same mass as protons. (Mass on Earth equals weight, but weight depends on the gravitational pull while mass does not. Your weight on the Moon would be much less than on Earth but your mass would be the same because it depends only on the amount of matter making up your body.) All atoms except normal hydrogen have neutrons, but the numbers vary. They occur in the tiny atomic nucleus tightly combined with the protons, and the mass of an atom is almost entirely due to these two components. Electrons, which have only one two-thousandth the mass of a proton, are outside the nucleus. Figure 1.1 shows diagrams of a hydrogen atom and a helium atom.

The hydrogen atom, with one proton, has one unit of mass; all the rest of the elements have masses that are multiples of this. Since the number of neutrons in an element can vary, atoms of a given element may have a variety of atomic masses. These are known as *isotopes*. Deuterium is the name given to hydrogen with a single neutron. It is also known as heavy hydrogen, giving rise to heavy water. Neutrons allow the positively charged protons to pack together in the nucleus, with larger nuclei needing more neutrons to hold them together. When a nucleus is very large the neutrons struggle to overcome the repulsion between the protons; this explains why a

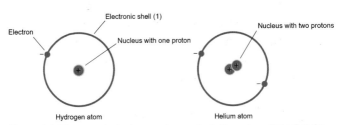

Figure 1.1: Diagram of a hydrogen atom and a helium atom. Note that the circles represent the pathway the electrons take as they move around the nucleus (the electron shells). The shells are not physical structures; atoms have no visible boundaries but have an electronic barrier which prevents them invading each other's space (see text).

particular isotope of uranium (235) can be used in nuclear fission bombs in which the atom splits into two smaller ones with the release of tremendous amounts of energy.

Where did atoms other than hydrogen come from?

Atoms are made in the stars. On Earth they are eternally unchanged and unchangeable, with the minor exceptions of the decay of radioactive atoms into other elements and a few alterations made by atomic physicists. Production of new elements from hydrogen is made by fusions between nuclei of atoms, and this can only occur at the high temperatures and pressures inside stars. Stars form from clouds of hydrogen which increase in size as gravitational attraction captures more hydrogen atoms. The tremendous gravitational pressure at the centre of the mass heats up the interior until it reaches the critical point at which protons of hydrogen atoms are able to fuse together, despite electrical repulsions, to produce helium, and in the process liberate tremendous amounts of energy. In other words, at this point the star lights up.

The process does not stop at production of helium. In stars like our Sun, further nuclear fusions between atoms produce elements up to atomic number 26, which is iron, but the process stops there. Further conversion of iron to elements with higher atomic numbers requires much higher temperatures than occur in the Sun. There is only one known type of event in the universe where this occurs: in the explosion as a supernova of a star ten times as massive as the Sun. In stars, the heat production by fusion of hydrogen and other elements counteracts the enormous internal gravitational pressures, but as the star runs out of fuel gravity gets the upper hand and the star contracts. In a small star like our Sun, repulsion between the atoms will eventually counter the effect of gravity, but a more massive star can contract further, developing indescribably high temperatures and pressures at the centre that result in the external layers being blasted into space in a gigantic explosion. For a few weeks the exploding star may radiate as much energy as the entire galaxy or as much as the Sun will radiate in its entire life. During this process, the elements heavier than iron, up to uranium the heaviest, are produced.

The series of nuclear reactions which produce all the elements have been worked out in detail. The debris from the explosion was scattered into the clouds of hydrogen gas. Ultimately new stars are produced by gravitational attraction; one such is our Sun, accompanied by its planets, one of which of course is the Earth. This improbable-sounding series of events was the origin of all the 92 elements on Earth. Fred Hoyle, inventor of the term 'Big Bang', played a prominent part in this epic piece of research. He was a brilliant physicist who also became well known for his science fiction novel *The Black Cloud*.

The structure of atoms

When scientists realised that atoms contained positively and negatively charged components, they also realised that atoms had a structure that had to be explained.

Initially there was the so-called plum pudding atomic model in which sub-atomic components were dotted around the atom in some undefined medium rather like fruit in a plum pudding.

A major advance in understanding the structure of the atom was made by Ernest Rutherford, a New Zealander. After graduating in mathematics and physics he was awarded an 1851 Exhibition Scholarship. These scholarships were inspired by Prince Albert and funded by the profits of the Crystal Palace industrial exhibition of 1851 to bring young scientists from the Commonwealth (then the Empire) to England. Rutherford worked in Cambridge as a research student under J.J. Thomson, who discovered the electron in 1897, for which he was awarded the Nobel Prize. Rutherford moved to Montreal University, then became head of physics in Manchester University and finally was head of the famous Cavendish Laboratory in Cambridge. In photographs of him in the Cavendish period in his untidy tweed suit, he looks rather like an amiable farmer.

Rutherford had great insight into what experiments to do and in interpreting the results, and was one of the most important pioneers in the field. He received a Nobel Prize in 1908 for his work on α-particles, which are ejected from radium atoms when they undergo radioactive decay. They are, on atomic scales of things, fairly massive missiles consisting of two protons and two neutrons. Rutherford arranged for a colleague, Hans Geiger, and a student, Samuel Marsden, to fire these at an extremely thin foil of gold. Behind the foil was a fluorescent screen, which gives a flash of light when a particle hits it. The α-particles went straight through the foil without hindrance. At least most of them did, but a few were deflected strongly and some bounced right back. This was an amazing result. As Rutherford described it at the time, it was like firing a 15-inch shell at a sheet of tissue paper and it bouncing right back. The result led to the concept of the 'Rutherford atom' in which the atoms are almost entirely empty space with the protons and neutrons collected together into a tiny nucleus and the electrons circling around outside. The experimental result was due to an α-particle (which is positively charged due to the protons) by chance encountering the positively charged nucleus of a gold atom and being repelled or deflected sideways from it.

This experiment showed that the nucleus occupies a minute fraction of the space inside an atom but it has essentially all of the atom's mass. A spoonful of protons would weigh a vast number of tonnes. Your body is made of atoms like everything else, so you might well have difficulty in coming to terms with the fact that it is almost entirely empty nothingness. So is Mount Everest, a steel girder, your house, the Earth, and anything you care to think of. The Rutherford model of an atom resembles the solar system with electrons (planets) circling around the central nucleus (the Sun). This model, however, conflicts with the fact that an electron circling the atomic nucleus should lose energy by radiation as it moves around the nucleus. Were this to happen the electron would quickly spiral into the nucleus and destroy the atom. The eternal existence of atoms depends on this never happening.

The problem was partly resolved by another of the great giants in atomic physics, Neils Bohr. He was head of his research institute in Copenhagen. He is described as a softly spoken, much liked, gentle person. He is often quoted as saying that predictions are difficult, especially into the future. Bohr joined Rutherford's laboratory, then in Manchester, and worked on the structure of the hydrogen atom, the smallest of all. Bohr's model recognised the difficulty in the Rutherford model and he postulated that the electron of a hydrogen atom was restricted to a defined zone or orbit around the nucleus, meaning that it couldn't crash into the nucleus. The orbit has no visible physical reality. Bohr's concept was important in that it correctly explained some properties of the hydrogen atom, but no classical model could give a correct explanation of why the circling electron did not crash into the nucleus. Were this to happen, matter would disappear.

The correct explanation came from quantum mechanics. In the 17th century, Isaac Newton, working in his rooms in Trinity College, Cambridge, had produced his laws of motion which explained the behaviour of macro objects such as planets moving around the Sun. His work was the basis of much of science for two centuries. The Newtonian view of the world was orderly and in accord with commonsense perceptions. All happenings had a cause and were predictable from the forces acting on systems. Gravitational attractions held the solar system together, eclipses were accurately predicted, and so on. It seemed that everything in the universe behaved according to his laws. The term 'clockwork universe' is sometimes used to describe it. It almost looked as if the science of physics was essentially complete with not much else remaining to be elucidated – a massive misapprehension to put it mildly.

Early in the twentieth century it was realised that, when it came to extremely small things like the subatomic constituents, the laws which so adequately explained the behaviour of everyday (macro) objects were really inadequate. When we enter this ultra-micro world it seems like an *Alice in Wonderland* Mad Hatter's tea party where impossible things are happening. They conflict with common sense, but they are experimentally proven to happen.

A new system of mathematics called quantum mechanics or quantum theory was developed to cope with the situation, and probably few, outside workers in the field, understand it. One of the most brilliant atomic physicists, Nobel Prize-winner Richard Feynman of New York (an eccentric, fun-loving character who was also a bongo drum performer), doubted whether anyone really understood it. Apparently no one understands why it works so well. Nevertheless quantum mechanics is now an indispensable part of physics and one of the most successful theories ever. If you would like to get a little more background to quantum mechanics, Box 1.1 gives some further reading. It won't make you understand the subject, but it might introduce you to its strangeness.

Arrangement of components in atoms

First, let us try to picture the hydrogen atom, the simplest of all. The atom is held together by the electrical attraction between the proton and the electron. Imagine

Box 1.1: Quantum mechanics

One of the strange things about the universe, which has no explanation, is that its nature can be predicted successfully by the language of mathematics. It is not clear why a person scribbling symbols on paper can deduce the most fundamental truths about the universe, but such is the case. It almost looks as if the universe is constructed on a mathematical basis. Quantum theory has been supremely successful. Its conclusions, though often starkly conflicting with human perceptions, have been verified by experiment whenever they have been tested.

The subject got its name originally from the work of Max Planck in Germany. Planck found that when a hot bar radiates heat, it doesn't come out as a continuous diffuse wave-like radiation as had always been assumed, but as discrete packets of energy which he called quanta. Planck was disturbed by the implications of his work for he realised it was likely to upset much of accepted physics. He tended to the view that his heat quanta had no fundamental physical reality but had a secondary explanation. However Einstein showed that light is also quantised, the energy being carried as discrete packages known as photons, but that it also travels as a wave. Light exists both as a wave and a particle; you cannot state which it is – it is both, a concept difficult to reconcile with commonsense predictions.

Curiously Einstein, who had done so much to lay the foundation of quantum mechanics with his work on quantised light, later fought a losing battle against the new theory, mainly because it was based on probabilities. He found it difficult to accept that in the

Box 1.1: (Continued)

quantum world things could happen without a cause but by chance. It is often quoted that he said he could not accept the idea that God played dice with the universe. Despite quantum mechanics becoming accepted as one of the most successful theories ever, Einstein never became reconciled with it.

Quantum theory predicted that electrons have dual morphology and in the 1920s it was realised that this was the explanation for the stability of atoms. The electrons exist both as compact bullet-like particles and also as a diffuse wave, rather like the ripples on a pond when a stone is thrown in. When waves are confined they can produce a 'stationary' or 'standing' wave form. The wave form gave a fundamental explanation of why the electron does not spiral into the nucleus; in the atomic structure the standing wave form cannot fall below a minimum energy level and therefore the electron cannot spiral into the nucleus. It's nice to have that settled; everything won't suddenly disappear.

Richard Feynman, already referred to, is widely regarded as a great genius in atomic physics. He was a great teacher and was aware of the mental block that facts that conflict with perceptions can cause in students. He warned them against tormenting themselves with futile questions that can never be answered because no one knows the answer. So he said, do not ask yourselves how atoms can be as they are. Physics tells us what they are but not how or why they can be like that. Best to take Feynman's advice. If he didn't understand it, you need not feel bad if you can't.

the atom to be a sphere with a diameter equal to the length of a football stadium. The nucleus would be about the size of a pea in the centre, and the minute electron circles around it at a distance. All the rest is empty nothingness. There is no physical boundary to the atom and yet it has a definite diameter, which can be measured. If you try to squeeze two atoms together you can get them to touch, but no further. They can't invade each other's space. There is an invisible boundary because the electrons of the two atoms repel each other. Like charges (+/+ and −/−) repel, and unlike charges (+/−) attract. This is why when you stand on a floor you don't go through its empty space. The atoms of your feet and those of the floor repel each other. It is electrical repulsion between the two that supports you.

The electrons of atoms are confined to specific zones known as electron shells. There is no physical structure to a shell (they represent different energy levels of the electrons in them). Each shell can accommodate only a maximum fixed number of electrons. The single electron of hydrogen needs only one shell to accommodate it. Heavier atoms require multiple shells, and these are arranged one outside the other somewhat like the leaves of an onion. They are numbered 1, 2, 3, and so on and fill progressively as the number of electrons to be accommodated increases. Most shells can accommodate eight electrons but hydrogen's shell takes only two.

Electron shells seem a long way from food production, but they play a crucial role in harnessing the energy of sunlight in photosynthesis, on which all food on Earth depends. If you would like to see more on this, see Box 2.1 (page 30).

What are molecules?

In living organisms, free atoms, by which we mean single atoms not connected to others, have very little role in the processes of living. Atoms in general are very sociable; they readily join up with other atoms by forming bonds to form clusters of atoms called molecules. Tens of thousands of different molecules are known, each having a specific collection of atoms joined together in a specific way. A molecule of, say, glucose is identical to all other molecules of glucose, whether it was made by a plant or in a chemistry laboratory.

The bonds between the atoms of molecules are quite strong so some can exist unchanged for long periods. Almost all familiar things around you are molecules. Some are very small, such as the hydrogen molecule consisting of two hydrogen atoms joined together (Fig. 1.2). This is the form that normally occurs because hydrogen atoms react immediately together to form molecules. Oxygen and nitrogen in the air similarly join together in pairs. Molecules of all sizes exist. Sugars, alcohol, vitamins, fats and suchlike are small molecules consisting of at most a few dozen atoms joined together (Fig. 1.3). Giant molecules also exist, such as starch, cellulose, proteins (Chapter 4) and DNA (Chapter 5), to give a few examples. These very large molecules are formed by joining together huge numbers of small molecules. Starch, for example, is formed from hundreds of small glucose molecules.

These are the stuff of life. Life is a molecular process in which molecules swap atoms with one another to form different molecules. There is constant chemical change going on both in living cells and in many non-living

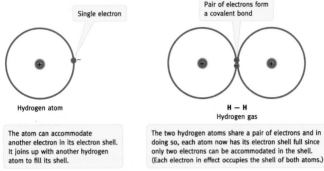

Single electron

Pair of electrons form a covalent bond

Hydrogen atom

H — H
Hydrogen gas

The atom can accommodate another electron in its electron shell. It joins up with another hydrogen atom to fill its shell.

The two hydrogen atoms share a pair of electrons and in doing so, each atom now has its electron shell full since only two electrons can be accommodated in the shell. (Each electron in effect occupies the shell of both atoms.)

Figure 1.2: Formation of a covalent bond between two hydrogen atoms to form a hydrogen molecule. Two electrons, one from each atom, form an electron pair; they both attract the nucleus of both atoms and this holds the two atoms together. Heat energy is released by the reaction.

materials. The activities between atoms and between molecules are called chemical reactions. It is how they react with one another.

An interesting question is why atoms are engaged in this restless activity. Why don't they just exist, being content with being an atom and doing nothing at all?

Figure 1.3: Glucose is a small molecule consisting of 6 carbon atoms, 6 oxygen atoms and 12 hydrogen atoms.

Oddly enough there is a small class of atoms that do exactly this. They are totally unsociable; they have nothing to do with all the other atoms. They do not react with any of them. They are chemically inert. They are all called noble gases, and include some that you may have heard of. The smallest is helium, then neon and krypton, and there are others.

Chemical reactions occur because non-noble atoms – the plebeian lot – want to become noble (to attribute human desires to them). But noble gases, like all atoms other than hydrogen, are made in the stars – they have to be born noble. So there is no way the other atoms can become noble. Instead they try to be as similar to noble gases as possible. How do they do this? They indulge in chemical reactions in which atoms join together to form specific collections known as molecules held together by chemical bonds. This of course does not make them noble, but they form structures in which each atom in the molecule mimics a noble gas.

So far I have dealt with the social aspirations of atoms, but not with why they want, as it were, to become noble. What is it about noble gases that makes other atoms want to resemble them? It is that the noble gas atoms have the most stable electron structure possible, which equates to having the lowest energy level. Everything in the universe tends to seek stability. Water falls downhill to a lower potential energy level; rocks fall off cliffs. It's just the way the universe is. Mountains crumble to dust; so do houses left to themselves. The decayed state with all their components spread haphazardly on the ground is at a lower energy level and is more stable than the intact mountain or house. Put in another way, the universe 'wants' to achieve the maximum possible disorder. Everything in it tends towards this. It is just the way the universe is. The drive towards maximum stability of everything is relentless and irresistible. There is a term in science that quantitates the degree of disorder in any system. It is called entropy. Entropy, as one writer put it, is a measure of the degree of mixedupness.

One of the natural laws of the universe, the all-important *second law of thermodynamics* (about which more in Chapter 2) states that nothing whatsoever can happen in the universe unless that happening makes a contribution to increasing the total entropy of the universe. The drive for increased entropy appears to be the reason for everything that happens in the universe. If infinite entropy is ever achieved, there will be a silent, dark, cold universe in which nothing whatsoever can happen, since when entropy is infinite you cannot increase it and therefore no happenings are possible. On current thinking that seems to be the inevitable end, but it's a long time away.

Atoms have the same drive to maximum stability as anything else. Noble gases all have something in common about them that makes them chemically inert and to have the maximum stability. It is that their outermost electron shell is full. Shell number 1, which is the outer shell of hydrogen and helium, can accommodate only two electrons, but for other electron shells the number is more usually eight. Atoms cannot alter the number of electrons they have, but chemical reactions can result in them mimicking noble gases in this respect by giving them full electron shells. It is always the outer electron shells that are involved in atoms reacting together. This is not surprising since for reactions to occur the atoms have to be in contact, and obviously it is the outer shells which are in contact.

There is a spectacular example in the stability difference between hydrogen and helium (Fig. 1.1). Hydrogen has a single electron. It is extremely chemically reactive. Helium has just one more electron, giving it a full outer shell, and this has a dramatic effect. Helium is a noble gas. It won't burn and it is chemically inert. For this reason airships, for safety, are now filled with helium rather than hydrogen. The use of helium was introduced following the *Hindenburg* disaster. On 6 May 1937, this giant airship was landing at Lakehurst, New Jersey, having just completed a crossing of the Atlantic, when it burst into flame and killed 36 passengers.

The chemical properties of the elements are determined by the electrons

How do chemical reactions cause atoms to mimic noble gases? When two reacting hydrogen atoms bump together, one electron of each goes to form a pair. Each attracts the proton on both atoms and this holds the pair together, forming a chemical bond between the two. It is a strong bond known as a covalent bond (pronounced coe vailent). Valent means strong.

In the case of bond formation between two hydrogen atoms (Fig. 1.2), the electron pair occupies the electron orbits of *both* atoms at the same time, a difficult thing to picture. You can think of the electron orbits of both atoms being fused into one which encircles both protons. Since the orbits accommodate only two electrons, both can be thought of as being full, and in this sense they mimic the noble gas configuration. Hydrogen molecules do not have a true noble gas structure, but the energy level of the hydrogen molecules is less than that of hydrogen atoms.

An element may form more than one covalent bond to fill its outside (valence) shell. In the case of hydrogen it needs only a single bond, since the shell only can hold two electrons – one from each of the two atoms. Many atoms whose outside shell can take eight electrons can form more than one such bond. They can indulge in covalent arrangements with other atoms until this number is reached. As already implied, the two paired electrons in a covalent bond are 'counted' as contributing to the valence shell of *both* the atoms involved. We can illustrate this with the carbon atom shown in Fig. 1.4. It has four electrons in its outer shell so carbon can form four covalent bonds with other atoms to completely fill its quota of eight electrons. Figure 1.5 shows the structure of methane, in which a carbon atom is attached to four hydrogen atoms. Both of the electrons in the covalent bonds are 'counted' for all the atoms. This gives the carbon atom an *apparent* full outer shell, mimicking a noble gas, and each hydrogen has a helium-like electron structure with two electrons filling its outer shell. It is a win–win situation for both carbon and hydrogen atoms. Methane does not have noble gas inertness but its formation reduces the energy level of the carbon and hydrogen atoms.

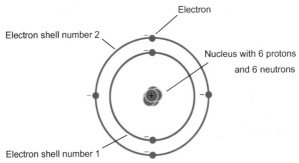

Figure 1.4: The electronic configuration of a carbon atom.

Covalent bond formation is of central importance in the struggle of atoms to achieve full outer electron shells. But there is an alternative method for some atoms. I will give an example of this which illustrates in a most dramatic way the effect of the electronic configuration on the chemical activities of atoms. It concerns table salt or sodium chloride, which is naturally in our foods so we eat it every day.

Sodium chloride (NaCl) consists of the elements sodium and chlorine. Sodium in pure metallic form is a dangerous substance that reacts violently with water, so much so that a piece dropped into water turns white hot due to the heat released by its reaction with the water. Chlorine is the poison gas used in World War I that had such terrible effects on soldiers exposed to it. And yet you eat sodium chloride or table salt every day. The explanation of this paradox is clear from looking at the electronic configurations of sodium and chlorine, shown in Fig. 1.6 in a simplified form. (Since only the outer shell is concerned in the chemical properties of the atom for this purpose it is not necessary to show the underlying shells.)

(a)

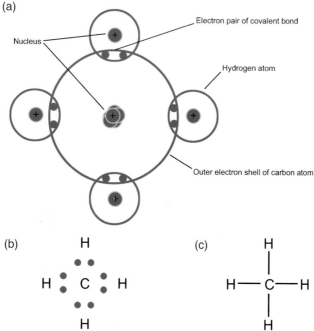

Figure 1.5: Three ways of representing the structure of methane.
(a) Electronic configuration of methane consisting of four hydrogen atoms
covalently attached to a carbon atom. Note that for simplicity only the outer
electron shell of the carbon atom is shown, since this is the only shell of
atoms involved in chemical reactions. (b) A Lewis structure in which each
shared outer shell electron is represented by a dot. (c) The usual way of
representing chemical structures. In this the covalent bonds are
represented by simple lines.

Shell number 3 of sodium has a single electron. If this
electron could be got rid of, the shell would disappear.
The underlying shell 2 would then be the outer one. This
is already full with eight electrons. The atom would now
have a noble gas configuration the same as that of the
noble gas neon. But how can the sodium get rid of the

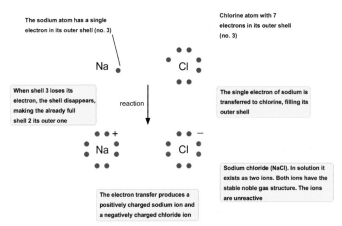

Figure 1.6: Reaction between sodium and chlorine. In this simplified diagram only the outer electron shells of the atoms are shown.

unwanted electron? It hands it over to another atom, chlorine, which is just as keen, as it were, to acquire an extra electron, as sodium is to lose it. The third (outer) shell of chlorine has seven electrons. If it gained one more it also would have a full outer shell and have the noble gas configuration of argon. Sodium and chlorine react in this way. It is a real win–win situation for the two atoms in their striving for noble gas status.

All atoms have equal numbers of positive and negative electrical charges that exactly neutralise each other so the atom has no net charge. However, when sodium hands over an electron it loses one negative charge and so becomes positively charged. By the same token, with the acceptance of one electron to fill shell 3, chlorine acquires a negative charge. Atoms with a charge are known as *ions*, so sodium chloride consists of a positive sodium ion and a negatively charged chloride ion. These ions have a genuine noble gas configuration with a full octet of electrons in their outer shells, and are extremely stable. This is not the creation of a new noble atom, which is impossible on Earth. The sodium ion still has the same number of protons as the original sodium atom, and the chloride ion has the number of protons of a chlorine atom. It is the proton number which defines what an element is. It's just that in the ionic form sodium and chlorine are unreactive. In solid form the two ions are electrically attracted to each other, while in solution they separate into free ions which can conduct electric current. Such compounds are known as electrolytes, and are commonly measured in doctors' blood tests.

In the history of chemistry the discovery of ions was a landmark achievement, but its author, Svante Arrhenius, a Swedish PhD student in chemistry, almost had his thesis rejected. At the time everyone knew that sodium reacts violently with water and that chlorine is corrosive. Therefore it was reasoned that sodium chloride could not separate into sodium and chlorine moieties. It was not appreciated that sodium and chloride ions behave differently from sodium and chlorine atoms, since ions were not previously known to exist. Arrhenius was finally awarded the lowest level of the doctorate. Later he was awarded a Nobel Prize for the work.

With that background on the nature of atoms we can proceed to discuss the problems facing the establishment of life.

2

The problem of how to supply energy to living processes

Life depends on a supply of energy for both physical work and the chemical work of building up the substance of the living matter. We get the energy from the food that we eat and ultimately that energy comes from the Sun, energy that photosynthesis traps for food production. In parenthesis, the trapping of light energy by plants depends on the electron shells of certain atoms (Box 2.1).

There are many energy-related problems which life had to solve.

How is energy liberated from different types of food and in what form? The answer to that question covers a

Box 2.1: Release of energy in the Sun and its trapping in photosynthesis

The energy in our food originated from sunlight, which in turn is derived from the fusion of hydrogen atoms in the Sun to form helium. It was early realised that ordinary burning of hydrogen as occurs on Earth could not be the source of the tremendous output of energy from the Sun. Einstein showed that energy and mass are interconvertible; this relationship he stated in the world's most famous equation $E = mc^2$, where E equals energy, m equals mass and c equals the speed of light (300 000 km per second). When you square the value of c you get an enormous figure, and the energy released from hydrogen fusion is (literally) astronomic. The Sun converts about 600 million tonnes of hydrogen per second into helium. It is middle-aged after 4.5 billion years of shining, and still has about another 5 billion years to go before its hydrogen fuel is exhausted. The prodigious rate of hydrogen usage need not cause feelings of panic about its running out.

The energy in sunshine travels to Earth as packets called photons. (Light also exists as a wave; it has dual wave–particle morphology – Box 1.1.) The light is absorbed by molecules of chlorophyll, contained in the membrane-enclosed structures in plant cells called chloroplasts. When a photon of the correct energy hits an electron of an atom in the chlorophyll it causes it to jump into a higher energy orbit. This has profound relevance to you and me because it is the basis of photosynthesis. The leaf has mechanisms (which have been elucidated) to extract the energised electron and use it to convert carbon dioxide (CO_2) and water to form sugars. The chlorophyll replaces the electron it

Box 2.1: (Continued)

used by extracting electrons from water, a process that releases oxygen. This supplies all the food on Earth and provides the oxygen organisms need to use it. It is a beautiful self-contained system.

massive amount of knowledge that has been discovered under the heading of metabolism (see page 36).

How is the liberated energy used to drive all the many processes essential for life to exist?

The scientists most associated with the answers to these two questions, Peter Mitchell, Fritz Lipmann and Hans Krebs, all received Nobel Prizes for their work. We were privileged to know all three in the 1950s.

Most of us think we know what energy is, for it is such a familiar term and is an everyday topic. But what is energy? You cannot see it or touch it, although we handle sources of it such as food and fuel. A useful everyday description is that energy is what makes things happen.

The scientific study of energy has the somewhat daunting name of thermodynamics, which simply means the movements and transformations of energy, heat being the most familiar form. There are laws of thermodynamics that describe how energy behaves. The development of the subject had its heyday at the time when steam engines, which depend on heat movement, were developing.

The laws of thermodynamics are of practical value for they define what can and cannot happen. If I give, as an example, that they make it clear why water cannot on its own run uphill, you won't be impressed because you knew that anyway. But the laws apply to things that can't be seen and where it is not at all obvious what can or cannot happen without energy being supplied to drive it. If you take chemical reactions, including those that constitute life, some can happen on their own and some can't. If a chemical reaction involves an increase in energy level it cannot occur spontaneously, but just as water can be pumped uphill, so 'uphill' chemical reactions can be made to go if energy is supplied. Life depends on uphill reactions to make its living substance. Cells take in small molecules from inanimate matter and convert them into the large high-energy molecules essential for life. The joining together of the small molecules needs energy to be supplied.

To continue the analogy, the energy for pumping water uphill often comes from the burning of coal, part of the released energy being converted to the electrical energy that drives the pump. The energy for driving the uphill chemical reactions of life is released when various types of food molecules are oxidised, that is, broken down through reactions with oxygen. In both cases the required energy comes from another energy source; coal and food respectively in the examples given. The big problem for the establishment of life was in what form the energy could be supplied for such chemical and physical work.

This preliminary statement has covered the essentials of the first law of thermodynamics, which states that energy cannot be created or destroyed, but different forms of energy can be converted into one another. The total amount of energy in the universe remains constant and the first law is, for this reason, also described as the law of conservation of energy.

If energy cannot be destroyed, why is the world so worried about it running out? We need to distinguish between energy that is available to do work and non-usable energy that cannot drive useful work. For example, the heat in a car engine block cannot be used to propel the car but it is still energy. Heat released from the oxidation of food in your body keeps you warm but it cannot drive your life processes. It cannot do work, either physical or chemical.

Everything contains heat energy. Atoms are not stationary; in liquids and gases they jiggle around in random motion, their velocity increasing with temperature. Even in solids, atoms vibrate and move to some extent. Molecules, which are groups of atoms joined together, also are in constant motion. Nothing is still.

Energy, as already implied, exists in various forms apart from heat. A moving object has kinetic energy (from the Greek word for moving). Kinetic energy is converted into heat energy by friction with air molecules. High-temperature heat can be converted into electrical energy via a steam turbine. There is also potential energy. Imagine a rock perched up on a cliff edge. It has gravitational potential energy. If it topples over the edge this is transformed into kinetic energy, part of which it will lose as heat generated by friction with air molecules as it falls, and all of it when it crashes at the bottom. You might regard it as stored energy; the higher the rock the greater the amount of stored energy it has. A rock on a ledge halfway down the cliff has half the gravitational potential energy of one at the top.

Most importantly, so far as life is concerned, there is also chemical potential energy. Oil, coal and wood have high levels that are released by burning, while water has none. It won't burn. The same is true of biological molecules. Sugars and fats have high calorific values, which means high levels of potential chemical energy, while amino acids (from protein digestion) have less. Different molecules have their own level of potential energy. If these molecules are oxidised, their chemical potential energy is released. This is what drives all of your living processes. It is why you need to breathe. The oxygen we breathe in is essential for food oxidation. None of this, however, explains how life couples the release of energy from food into driving the essential uphill reactions of the living process or how muscles use the released energy for physical work. But before we get to that problem we need to look further at the laws of the universe governing energy.

In Chapter 1 (page 22) I discussed entropy and its importance in explaining why things happen in the universe. The second law of thermodynamics is of central importance. It formally states that *every happening must increase the total entropy of the universe.* Another way of expressing it is that nothing can happen unless the happening is energetically downhill. Chemical reactions can get the nod of approval from the second law only if they result in the products of the reaction having less energy than the starting reactants. In other words, chemical reactions proceed only if they liberate energy, most commonly in the form of heat. Liberated heat causes molecules in the air to jiggle around more vigorously, and this increases entropy. No process can therefore ever be 100% efficient energetically since only the part left over from increasing entropy is available to do work. (The useful leftover part has the potentially confusing name of 'free energy'. This does not mean 'free' as in 'for nothing' but rather 'free' as 'available to do work'.) As discussed in more detail in Chapter 1, the universe is driven by its tendency to achieve infinite entropy, at which point, if achieved, nothing whatsoever can happen since entropy cannot be further increased.

What do living organisms need energy for?

As indicated above, to grow and reproduce, organisms have to make their own substance. They do this using small components from the diet which they join together by chemical bonds to form the larger molecules of which living matter is composed. Examples are the assembly of amino acids in large numbers to make huge protein molecules (described in Chapter 4). Sugars are linked or polymerised to form starch in plants, and a similar process operates in animals to store energy in muscles as a starch-like polymer called glycogen. This chemical synthesis activity is going on constantly in thousands of reactions producing the many different structures of life.

The large molecules have a higher energy level than the small components from which they are formed. The synthetic processes are energetically uphill and therefore energy must be supplied to drive them. This comes from oxidising high calorific foods such as fat and sugars to release their chemical potential energy. Life, looked at broadly, consists of the uphill chemistry of synthetic work driven by downhill chemistry of oxidation of high calorific food to supply the needed energy. It's like a pulley system, with high calorific foods in a basket high up on one side of the rope and small molecules of low energy in a basket at the bottom on the other end. As the high calorific foods drop down the energy cliff (by being oxidised) they pull up the low basket to the high energy level. Put in more scientific terms, the breakdown of foods is called catabolism and the building up of large molecules is called anabolism, the two together being metabolism. Catabolism drives anabolism.

One might get the impression that living cells defy the second law of thermodynamics. They assemble their substance from small molecules randomly scattered in the environment or from the digestion products of food. These have a high entropy level. The living cell is highly organised with a much lower entropy than the starting materials. This might suggest that, after all, life is a magical process that defies the laws of the universe. This is not the case, because the entropy *increase* caused by the breakdown of food such as glucose to provide the energy for cell growth is greater than the entropy *decrease* caused by assembly of the cell. The total process of the assembly of living matter plus the oxidation of food therefore increases the total entropy of the universe, and it is the total to which the second law applies, not to individual components of the process. There is a parallel to this in the building of houses. The bricks, wood, tiles and steel components are dumped on the building site any old how – they have a high entropy level. The builder expends a lot of energy transforming them into a highly organised house which has a lower entropy level. The energy for this comes from food oxidation in the builder's body, which increases entropy more than the assembly of the house lowers it. Building the living cell and the house both comply with the second law, as indeed everything must. There is no special pleading when it comes to the laws, and there are no exceptions. Knowledge of them would have saved much time and effort in the past of people trying to design perpetual motion machines that could run forever.

The universal energy currency unit of all life forms

We now come to one of the major problems that life had to cope with. How do you connect the release of energy from oxidation of foods like glucose and fats and other things to drive energy-requiring processes such as chemical synthesis and muscle contraction? It is, in principle, similar to the problem of how you connect the burning of petrol to the propulsion of a car. We know that this is done by burning the petrol in cylinders and connecting the resultant piston movement mechanically to the wheels.

Connecting energy release from food oxidation to do work in living organisms is vastly more complicated. The body cannot use heat liberation to drive things; the energy must be in a form that it can couple to its living processes. The problem is further complicated because a variety of different foodstuffs are used as energy sources and the energy released has to be coupled to a very large number of different tasks: it is needed to make muscles contract, for impulses to move along the nerves, for vision to occur, to drive the formation of thousands of different molecules essential to life by uphill chemical reactions, and to transport molecules around the cell as needed, to cite just a small fraction of the total. The energy has to be supplied by the oxidation of different foods – fat, sugars, proteins, even alcohol. It is not the purpose of this book to go into all the enzymic reactions involved in answering this question, but it may be of some interest to know a little bit about the man who in 1937 unravelled the sequence of biochemical reactions involved. This sequence is known as the citric acid cycle, often called the Krebs cycle.

Hans Adolf Krebs is best known for his identification of two important metabolic cycles: the urea cycle, and the citric acid cycle or Krebs cycle. He received the Nobel Prize in 1953 for his work on the Krebs cycle, sharing the Prize with Fritz Lipmann (whose work is discussed below). Krebs emigrated to England in 1933, and in 1954 succeeded Sir Rudolf Peters in the Chair of Biochemistry at Oxford, when Peters retired. (Some of Peters' work is discussed in Chapter 3.)

Krebs had a childlike sense of humour which we often observed during our time overlapping at Oxford. At one of the Symposia organised by ex-colleagues and students of Fritz Lipmann (the fourth such event, at

Berlin in 1974), during one of the coffee breaks we were chatting with Krebs. With a beatific smile, he held up a biscuit and remarked 'Feeding the Krebs cycle'.

The Krebs cycle is the key sequence of metabolic reactions that produce energy when dietary foods are broken down in cells. It consists of a circular series of eight biochemical reactions. Through each passage of the cycle four hydrogen ions (H^+) are generated. The ions are actively pumped across the mitochondrial membrane as we shall see (Fig. 2.3) by a connected process called the electron transport chain (also called the respiratory chain), resulting in a net accumulation of H^+ from the inside to the outside of the membrane.

How can you connect energy release from a variety of sources to drive a multitude of different energy-requiring tasks? It seems to be an impossible job.

There is an analogous problem in human societies. We get energy from burning coal, wood, and oil. How can you couple the burning of the different fuels to each of the many different uses? The solution is mainly to transform the energy from all of them into electricity, which is transmitted to wherever it is needed. This is very flexible. Billions of years earlier, life hit on a similar strategy. It converts the energy from the oxidation of all types of sources into one single form of chemical energy, and this is used to drive everything in life. It is very flexible arrangement and majestic in concept.

All life on this planet is powered by the one relatively simple chemical molecule called ATP (pronounced as the three separate letters A-T-P), short for adenosine triphosphate. It is the universal energy currency of life, somewhat like the dollar is the monetary currency for much of the world. There are only a very few minor exceptions to this. All forms of life – bacteria, plants and animals – grow and reproduce using it. You synthesise the components of your body at the energetic expense of ATP. Whales swim using it. If you lift an arm, ATP powers the action; it powers your brain waves. Fireflies make flashes of light on it, electric fishes produce voltage shocks using it; in the past dinosaurs roamed on it; birds fly on it. Even viruses depend on it. Name any biological work and ATP supplies the energy for it. It is, in principle, such a simple concept that it may be difficult for you to accept that virtually all biological energy-requiring processes on this planet for billions of years have been driven by a single chemical molecule.

Adenosine triphosphate, despite its almost awe-inspiring role in life, can be obtained as an ordinary-looking white powder; it can be stored in the deep freeze. It can't be supplied to cells from the outside (Box 2.2); each has to make its own, for it cannot pass through the cell membrane into the cell where it is used. (It wouldn't do for such a vital thing to be able to leak from the cell.)

The body has to synthesise it continuously because there is very little reserve. This is why you cannot survive more than a short time without oxygen. When oxygen is not available, ATP synthesis stops. You might like to have a quick look at the remarkable molecule shown in Fig. 2.1 with its three phosphorus atoms (P). Next time you watch one of the magnificent wildlife programs on television, just reflect that all the action of birds flying, herds of animals rushing through the African plains or whales

Box 2.2: Mitochondria deliver energy right where it is needed

Mitochondria are circular or rod-like structures that exist inside nearly every cell in your body and create usable energy in the form of ATP.

The number of mitochondria in each cell, and where they are located relative to other key structures, varies hugely depending on the specific energy requirements and functions of the cell in question.

In sperm, mitochondria are found concentrated around the section of the tail immediately below the head, where the flicking movement for swimming is initiated. This means that ATP is manufactured exactly in the right position to deliver energy for cell propulsion.

In heart muscle cells, mitochondria form a lattice arrangement within the contractile units to provide energy for continual beating.

In skeletal muscle cells, mitochondria are found in pairs or lines along the length of the contractile units to supply energy for contraction when required for movement.

The body-building industry tries to cash in on the known role of ATP in creating energy by selling supplement powders that contain crystalline ATP. But because ATP cannot pass through the cellular and mitochondrial membranes to reach the contractile units inside muscle cells, use of such products will not assist your dream to become the next Arnold Schwarzenegger.

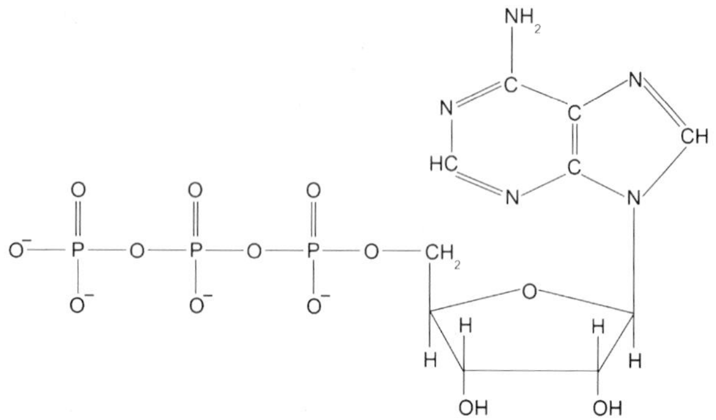

Figure 2.1: ATP, the molecule that powers everything in all forms of life. Just have a quick look at it – no need to understand it. The business end of the molecule is the chain of phosphorus (P) atoms.

thrashing about in the oceans is being driven by this molecule, which is not much bigger than a molecule of sugar. It's an almost incredible concept.

What makes the structure of ATP so suitable for its central role in all of life?

The element phosphorus plays a central role in the energy merry-go-round of life. Phosphorus in its commonest form exists as inorganic phosphate. Inorganic simply means that it is not attached to any carbon-containing compound. It is the convention in biochemistry to write inorganic phosphate as Pi – the 'i' stands for inorganic. It is in this form that you buy it as a garden fertiliser. In the context of biological energy, Pi has zero energy. It can be attached to other molecules by chemical bonds (page 21) and form what are called phosphate groups on molecules.

The phosphate groups are of two types. One is known as low-energy and the other as high-energy phosphate groups. The low-energy phosphate group we will write simply as –P and the high energy one as –P*. We can write a very simple diagram of ATP as Ad–P–P*–P*. Ad stands for a molecule called adenosine. The first –P attached to Ad is low-energy phosphate, but the next two are high-energy ones (for structural reasons I will not go into), meaning their energy can be extracted and used.

You can liken a living cell to a very busy city (Fig. 2.2). Each cell of the types found in animals and plants has several separate power stations, called mitochondria, scattered around pouring out ATP (cf. electricity). The more energy the cell requires, the greater the number of mitochondria (Box 2.2). A mitochondrion is a small structure surrounded by a membrane, found inside animal and plant cells (but not bacterial cells – see Fig. 2.2). A mitochondrion is about the same shape and size as a bacterial cell. This is no coincidence, because from an evolutionary viewpoint mitochondria originated millions of years ago by a bacterium-like organism becoming engulfed inside a precursor of modern animal and plant cells and becoming a permanent semi-independent passenger (see Box 8.2). This was an important landmark event in the evolution of cells. Mitochondria multiply inside the cell to meet the energy requirements of the cell, but have retained bacterial characteristics. When cells divide each daughter cell receives its complement of mitochondria.

The engulfed bacterial cell had developed the ability to use oxygen in the breakdown of food; the host cell that

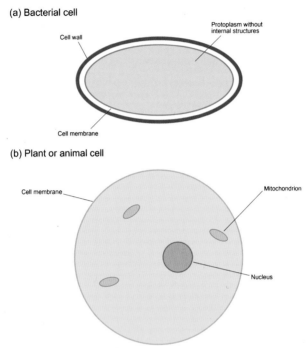

(a) Bacterial cell

Cell wall

Protoplasm without internal structures

Cell membrane

(b) Plant or animal cell

Cell membrane

Mitochondrion

Nucleus

Figure 2.2: Diagrams of (a) bacterial and (b) animal cells. Bacterial cells are about one thousand times smaller in volume than animal cells. They have no membrane-bound structures inside them. Animal and plant cells have many such inclusions, each with a different function. Among the membrane-bound inclusions, mitochondria are the sites of almost all of the ATP made by the cells. In bacteria the corresponding sites are in the cell membrane.

engulfed it (the precursor of animal and plant cells) at that time did not have this capability until it was supplied by the bacterium. A small amount of ATP can be produced from glucose in the absence of oxygen. Yeast

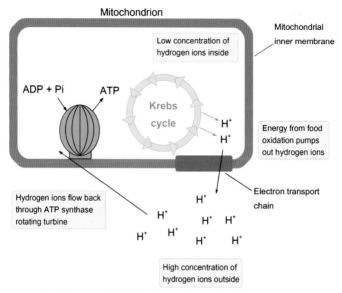

Figure 2.3: Diagram of Mitchell's theory of how the energy from food oxidation is used to produce ATP from ADP + phosphate. He postulated that the ions produced in the Krebs cycle are actively pumped out of the mitochondrion using the electron transport chain, resulting in a concentration gradient across the membrane, high outside and low inside. The hydrogen ions are allowed to flow back down the gradient into the mitochondrion through a turbine which is rotated. This leads to the production of ATP.

cells, for example, do this on a large scale by producing alcohol during beer making, but most of the chemical potential energy of the sugar is still left in the alcohol. Cells with mitochondria oxidise glucose to carbon dioxide, and ATP production is 16-fold higher than that in the absence of oxygen. This is a massive increase in efficiency of ATP production.

How is ATP used in the body to drive the synthesis of the large molecules of living matter?

Adenosine triphosphate is an ordinary chemical like thousands of others. It cannot do anything itself. Its high-energy phosphate groups must be transferred to other molecules when it functions as an energy supplier. Simplistically, you can think of it as transferring a packet of energy with the phosphate group. (Special enzymes do the transferring; enzymes are described in Chapter 3). Suppose life requires that two molecules we will call A and B have to be joined together to give a larger molecule, A–B; and that the energy level of A–B is 13 kJ higher than the energy level of A plus the energy level of B. Making A–B is an uphill chemical reaction. It cannot be formed without an energy input, just as water has to be pumped to flow uphill. The cell arranges that ATP breakdown is incorporated into the enzymic reaction which joins A and B together to make A–B (Box 2.3).

The importance of this chemical strategy of incorporating the energy of ATP into energy-requiring chemical reactions cannot be overstated. It might be described as one of the secrets of life. It applies to virtually everything in the body. There are hundreds of different enzymes transferring –P* groups to different molecules. As one example, when you raise your arm –P* groups are transferred from ATP molecules to your muscle proteins where they are then released as –Pi. This supplies the energy for the muscle to contract. The mitochondrial furnaces then immediately spring into action and convert the ADP + –Pi back to ATP again.

Box 2.3: How ATP drives energy-requiring reactions

An enzyme transfers a –P* group from ATP to A to give A–P*. A second enzyme now reacts with A–P* and adds B to A, displacing –P* which is released as –Pi (which has zero energy). We can summarise all this very simply as follows:

1st step: A + ATP → A–P* + ADP (ADP, adenosine diphosphate, is ATP which has lost one phosphate).
2nd step: A–P* + B → A–B + Pi

If you put these two reactions together as a summary of the whole process (and it is the whole process which counts so far as the second law is concerned) we get:

Overall: A + B +ATP → A–B +ADP + Pi

The energy supplied by ATP breakdown is 30 kJ but only 13 kJ is needed to join A and B together, so the overall process is 17 kJ downhill and will proceed to completion. It is completely in accord with the second law. For some chemical syntheses even more energy is needed, and for these a simple modification is used in the chemical reaction so that two high-energy phosphates are split off the ATP, giving a big energetic kick to the reaction.

The magnificent concept of life running on high-energy phosphate groups of ATP was put forward in 1941 by Fritz Lipmann working in a laboratory of the

Massachusetts General Hospital in Boston, which is part of the Harvard Medical School. Lipmann illustrated his concept as a rotating dynamo with ADP and Pi going in and ATP coming out, the dynamo being driven by food oxidation. It accurately outlined the pattern of energy utilisation of all life forms and his concepts are as true today as they were 50 years ago.

Lipmann was the archetype of a brilliant scientist totally absorbed in his work. He had an unworldly air, and was liked by all who knew him. When Bill was a member of Lipmann's laboratory he took lessons at a local driving school and obtained a licence. Lipmann, who had always wanted to learn to drive a car, was very impressed by this and asked him for the phone number of the school. His protective second-in-command, Dave Novelli, quietly said to Bill, 'For God's sake don't give it to him, he'll kill himself.' Bill managed to avoid doing so. Lipmann never got a driving licence, but in 1953 he did get a Nobel Prize. His phosphate bond energy concept had, and still has, a tremendous influence on research into the biochemistry of life.

A long research controversy: how is ATP produced in mitochondria?

Following Lipmann's publication, there was a great effort by many biochemists to work out exactly how ATP was synthesised from ADP and phosphate, that is, to discover how $-Pi$ of zero energy was converted to a high energy $-P^*$ group of ATP as a result of food oxidation. Many workers tried to solve this problem. For many years all attempts failed, to the great frustration of the workers; it was said at the time that conferences on the subject generated more heat than light and the research workers were often referred to as mitochondriacs.

The trouble was that it had been earlier established that a small amount of ATP is produced outside of mitochondria anaerobically (without oxygen); it occurs via the formation of a high-energy phosphate attached to a breakdown product of glucose which is then transferred to ADP by an enzyme. It was confidently believed therefore that such intermediates must be produced in mitochondria where most of the body's ATP is made. It was the only known mechanism for producing ATP. But all attempts to find the postulated intermediate failed.

The breakthrough which led to the solution came from Peter Mitchell in England. He started his biochemical career in Cambridge University where he was somewhat unusual in that when every other male researcher at that time wore white shirts, tweed jackets and grey trousers and had short-back-and-sides haircuts, Peter had long hair and colourful clothes. He had a brilliant mind and an expert knowledge of the more physical aspects of biochemistry. He left Cambridge for Edinburgh University; he and his wife took a holiday in north Devon where by chance he looked over a decaying huge Edwardian mansion with a small farm attached. He there encountered the agent selling it, who took his name and address. Sometime after returning to Edinburgh he was surprised to receive an offer from the agent with a sale price he decided to accept, and he withdrew from university employment.

A year or so was spent in renovating the mansion and he then set up a private research laboratory devoted to the problem of ATP synthesis, helped by his former Cambridge assistant Jennifer Moyle. At the time when we visited it he had a temporary laboratory in the huge entrance hall covered by a sheet of plastic to keep the dust out. Mitchell approached the problem in a unique way and in 1965 published a theory (called the chemi-osmotic theory) on how ATP is produced. It was so revolutionary that hardly anyone took it seriously and it was strongly opposed by many in the field. This was probably because the theory did not involve intermediates that everyone else had assumed were necessary, and it included concepts that were unfamiliar to many biochemists. Mitchell was a pleasant man with a good sense of humour but he was a worthy opponent in a debate. He responded to his critics saying that they were very imaginative people; they imagined intermediates that did not exist. The debate got unusually hot for academics. Seventeen years later, in 1978, Mitchell was awarded a Nobel Prize, for his theory had the advantage of being correct. In his address at the award ceremony in Stockholm he commented that he was gratified his theory had become accepted by his critics during their lifetime, because acceptance of a scientific idea was usually not due to conversion of opponents but to their dying off.

Mitchell's theory has a magnificent, majestic simplicity in explaining how all life has been driven for billions of years. His theory said that the energy derived from food oxidation is used to pump hydrogen ions (H^+)

from the inside to the outside of the mitochondrion through its thin membrane, using the electron transport chain. (Hydrogen ions are hydrogen atoms stripped of their electrons.) This results in a gradient of hydrogen ion concentration, so it is high on the outside and low on the inside of the cell. This is indicated by a fall in the outside pH. (pH is a measure of acidity familiar to many gardeners: low pH means high acidity; high pH means low acidity.)

Figure 2.3 gives the overall principle of Mitchell's theory, the simple basis of which is that gradients can do work. A water gradient can drive electric turbines or an air pressure gradient a windmill. A gradient of hydrogen ions can also do work if it is allowed to flow through a suitable device. Mitchell postulated that there are such devices in mitochondria. These have been found as protein complexes called ATP synthase located in the mitochondrial membrane, where they synthesise ATP. Bacteria have these in their cell membranes; since mitochondria originated in the evolutionary sense from bacteria, the presence of ATP synthase complexes in mitochondrial membranes all fits. These amazing molecular machines include a turbine-like protein structure which is rotated by the flow of hydrogen ions from the outside back into the mitochondrion. The rotating disc turns a shaft inside a spherical structure made up of six protein subunits. As the shaft rotates it contacts the subunits, causing a conformational (shape) change in the units. The energy for ATP synthesis comes from this; three ATP molecules are produced in a single complete rotation.

You have trillions of these ATP synthase complexes in your body. So have all living organisms on this planet that use oxygen – plants, animals and bacteria. They are virtually identical in all life forms.

It is quite awe-inspiring to reflect that virtually all life on this planet is driven by the creation of a small pH gradient across a microscopically thin membrane. Who could have imagined that such a simple device powers gigantic animals like whales and must have powered the dinosaurs? It is a beautiful concept.

The very important role that mitochondria play in generating energy to support life is illustrated by looking at the health problems of people suffering from mitochondrial disease (Box 2.4).

Box 2.4: Mitochondrial disease

The very important role that mitochondria play in generating energy to support life is illustrated by looking at the health problems in people suffering from mitochondrial disease. Mitochondrial disease is characterised by a failure somewhere along the chain of events that take place in the mitochondria to generate ATP for energy. The symptoms of mitochondrial disease occur not only because the cells involved become deprived of ATP for energy, but also because they start to accumulate unused molecules such as ATP precursors and oxygen. These molecules are then metabolised to make by-products that further damage the mitochondria and the cell. Without their source of

Box 2.4: (Continued)

energy and due to accumulation of by-products, the cells begin to function poorly and die.

The clinical symptoms of mitochondrial disease are extremely varied, depending on which cells of the body have defective mitochondria. Tissues in which cells have the highest energy demands and hence highest numbers of mitochondria, such as brain, heart, muscle, kidneys and liver, are most susceptible. One well-described form of the disease involves the optic nerve, the main nerve connecting the eye to the brain. Patients present with blurred vision, which progresses to clinical blindness. Clinical presentations due to abnormal mitochondria in other cell types include exercise intolerance, muscle weakness, seizures, dementia, stroke-like episodes, hearing loss, feeding and speech difficulties, and kidney failure.

There is currently no known cure for mitochondrial disease. Depending on which organs are affected, some patients do not survive childhood, whereas others can be managed clinically for many years.

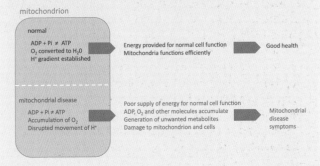

mitochondrion

normal

ADP + Pi ⇌ ATP
O_2 converted to H_2O
H^+ gradient established

Energy provided for normal cell function
Mitochondria functions efficiently

Good health

mitochondrial disease

ADP + Pi ⇌ ATP
Accumulation of O_2
Disrupted movement of H^+

Poor supply of energy for normal cell function
ADP, O_2 and other molecules accumulate
Generation of unwanted metabolites
Damage to mitochondrion and cells

Mitochondrial disease symptoms

In this chapter we have seen how life has coped with the laws of thermodynamics. But there is another fundamental problem which at first sight might seem to make a chemically based life mechanism impossible. Since we exist it obviously has been solved, but in a remarkable way as discussed in the next chapter.

3

Enzymes overcome the chemical reaction barrier problem

If, to use an impossible analogy, you were trying to plan life, at this stage you might be feeling optimistic after having worked out a strategy whereby life could cope with the laws of thermodynamics. But then you realise with horror that your scheme of basing life on a multitude of chemical reactions is impossible. Why? Because there is a barrier to the occurrence of chemical reactions. Something stops them even if they would not violate the requirements of the second law of thermodynamics. If you simply mixed the chemicals involved in life together, nothing at all would happen. The barrier stops them reacting. But if the reactions cannot occur then life based on chemical reactions is not possible.

Paradoxically, if the barrier were not there life would still be impossible, because without it all the possible chemical reactions would be over in an instant. So, the barrier appears to make life impossible and if there were no barrier life would still be impossible. It looks to be a difficult problem and a bleak outlook for the prospect of establishing life. To understand this situation we must look a little more closely at the nature of chemical reactions in a very simple way.

Consider a bowl of sugar on the table exposed to oxygen in the air. The thermodynamics are favourable to it reacting with the oxygen – being oxidised. If the sugar is oxidised to produce water and carbon dioxide, energy will be liberated as heat; the reaction is steeply downhill and would fulfil all the requirements of the second law. But the sugar will sit there on the table indefinitely, for decades if left alone, without anything happening to it. Something is stopping it burning or it being oxidised in any way. But if you ate the sugar it would quickly be involved in chemical reactions and be oxidised to carbon dioxide and water. What then is so different about the chemistry of living systems and that of inanimate objects? A century ago it was believed that it was due to the chemistry of living systems being a mysterious kind of super chemistry, quite different from that outside of living things and beyond the comprehension of science. This view was incorrect; living organisms *are* subject to the natural laws, but they have something that overcomes the barrier that inhibits chemical reactions and that made life possible.

The solution was the development of a new type of molecule, produced only by living systems, that allowed chemical reactions to occur at the high speeds required for life to exist. To explain this we must look at the nature of chemical reactions. To start with an analogy, imagine that a chemical reaction is equivalent to a rock falling off the edge of a cliff. It is not a bad analogy, because a chemical reaction is possible only if it involves a fall in energy of the total reactants – it must be a downhill reaction, just as a rock falling is a downhill process. The chemicals must fall over an energy cliff if reactions are to proceed.

To turn back to the rock analogy, if the rock is right at the edge of the cliff, it is likely that it will easily be pushed over by a very slight nudge; that is, by a small input of energy. But if the rock is a little way back from the edge with an upward slope between it and the edge (Fig. 3.1a), it will sit there indefinitely unless something pushes it up the hill to the very edge, when it will easily topple over (Fig. 3.1b). This requires a much larger input of energy and is a barrier to the rock falling over the edge.

This situation is comparable to a chemical reaction in that, as explained, it involves the chemical molecule falling down an energy cliff. Suppose the reaction is the conversion of molecule A to molecule B. This does not occur directly, in a single step (*not* A → B). Instead, A is first converted to an activated form (A*), which is then converted to the product B. So the reaction sequence is

(a)   (b)

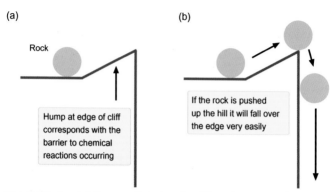

Figure 3.1: A rock falling over the edge of a cliff can be compared with a chemical reaction; (a) shows that a rock a little way from the edge of the cliff will not fall over because the slope is a barrier preventing this.

A → A* → B. The activated form A* is known as the *transition state of the molecule involved.* The transition state has a very brief existence, perhaps only a trillionth of a second, and then converts to B. It is at a higher energy level than A or B; it is an activated form, somewhere between A and B in structure, which readily converts to B. Energy is needed for formation of the transition state, and this is the barrier that stops chemical reactions occurring. Unless A is converted to the transition state the reaction cannot occur. For most reactions at low temperatures, the energy requirement to form the transition state is too great and the reaction does not occur spontaneously. So ordinary molecules such as sugar, paper, wood, and all the familiar things around you are quite stable. The situation is very similar to that of the rock falling off the cliff. A is equivalent to the rock behind the slope, and A* to the rock at the cliff edge. To continue with the analogy, in a chemical reaction molecule A has to be pushed up the hill by an input of energy to reach its transition state A*, and then it reacts to form B, liberating heat energy. The energy input is known as the activation energy.

This analysis applies to all chemical reactions whether in living or inanimate systems. It is just as well that the barrier is present, for if it was not there and ordinary chemicals reacted together instantly, everything would have been used up in reactions long ago so that no more reactions were possible and life could not have existed.

How then do you make reactions go? For inanimate things the solution is simple – you supply heat to raise molecules to their transition state. Chemists boil reactants, sometimes for hours; the heat supplies the necessary activation energy and reactions proceed happily. To oxidise petrol in a car cylinder the activation energy is supplied by the spark. To burn paper you supply it by a lighted match, and once the paper molecules start to burn the heat liberated maintains the reaction.

The problem is vastly more complicated in organisms. Life operates at low temperatures and in dilute watery solutions, and such methods are not applicable. To make life possible, a solution to the chemical reactivity barrier problem had to be found which does not require sparks or high temperatures. The problem for life, however, is much more complicated even than that. It must overcome the energy barrier for thousands of different chemical reactions. Each individual chemical has its own transition state, each of which has to be separately activated. When a chemist promotes a reaction by boiling, every possible reaction that the reactants can participate in will occur. The result is often a black soup from which your wanted product has to be isolated. This would not do for the reactions of life. A biochemical reaction in the body has to produce one product without other unwanted reactions occurring. A wanted chemical reaction in the body has to be promoted with great selectivity, activating only one chemical to its transition state. And this has to be done for thousands of different chemicals. It does seem, on the face of it, to be an absolutely impossible task. We do exist so the problem has been solved – by the development of a new type of molecule with properties that could well be the most remarkable in the universe. These molecules are called proteins, and are the subject of Chapter 4. One very special class of protein molecules, called enzymes, is the main subject of this chapter. The role of one enzyme is described in Box 3.1.

Box 3.1: Chemistry in your kitchen

Overcoming barriers to chemical reactions is a common occurrence in the kitchen: techniques that we use in our cooking today have developed over thousands of years, and were often initiated by chance or happy accident.

Over time and with physical pressure (i.e. kneading), wheat flour mixed with water and yeast changes as a result of the induced chemical interactions

Box 3.1: (Continued)

of its components. Water, gluten proteins, starch granules and gas created through fermentation create a complex three-dimensional structure that is fixed into place through cooking.

To make cakes, fast-acting chemical leaveners instead of yeasts are used to introduce gas. Products such as baking powder and bicarbonate of soda exploit a reaction between acidic and alkaline components of the cake ingredients to produce carbon dioxide. Cakes should be placed in the oven as soon as possible after the dry and wet ingredients are combined to ensure that the combination of heat and carbon dioxide production delivers a well-raised and light product.

Milk is a complex mixture of fat, protein, sugars, vitamins and water. Curdling occurs when the levels of acid in the milk rise: this allows the fatty elements to cluster together, and the protein molecules to clump into a three-dimensional network – the curds. A history of cheese-making reveals that about 5000 years ago people learned they could salt the solidified milk curds to make a long-lasting version of milk. At some point along the way they also realised that the cheese became more pliable and cohesive if the mixture included pieces of animal stomach. We know now that it was the presence of an enzyme (rennet) that was derived from the stomach cells that made the difference. Rennet alters the chemical nature of a specific milk protein and allows the molecules to link together to form a continuous solid curd. These days most rennet is not collected from cow stomachs, but made in the laboratory.

Enzymes bring about all the reactions of life

Enzymes are the biological catalysts that cause
reactions to occur in living organisms which otherwise
would only proceed, if at all, at negligible rates. A
catalyst accelerates the rates of a chemical reaction
without itself being changed. Enzymes are very large
molecules. A typical enzyme would be about a
thousand times larger than a molecule of glucose. The
molecule that an enzyme attacks (causes a reaction in)
is called the *enzyme substrate*; in most cases it is very
small in comparison with the enzyme itself. The
enzyme must bind to its substrate to bring about the
reaction. In the cell all chemical reactions occur in
molecules bound to their appropriate enzymes (Fig.
3.2). After binding of their substrates, the reaction
occurs. The product no longer binds to the enzyme and
leaves it, and the enzyme is then ready to receive the
next substrate molecule. Put in abbreviated form the
sequence is:

$$E + S \rightarrow ES \rightarrow EP \rightarrow E + P$$

(E = enzyme, S = substrate, P = product. The actual
reaction occurs in the conversion of ES, the enzyme–
substrate complex, to EP, the enzyme–product complex).

How does an enzyme bind to its correct substrate?
Each enzyme has on its surface a small patch called the
catalytic, or active, site. This is part of the enzyme
protein but the atoms in the patch are arranged so that its
surface has a shape complementary to the shape of its
substrate molecule, meaning that they can fit together

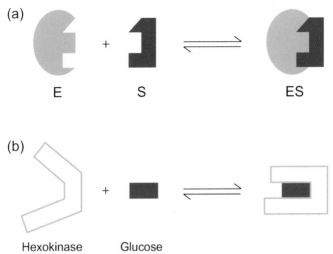

Hexokinase   Glucose

Figure 3.2: This diagram illustrates the principle of how an enzyme combines with its substrate molecule. The substrate has a shape which precisely fits into the catalytic site of the enzyme by weak chemical bonds. Then the reaction occurs and the products leave the site, which is then free to receive another substrate molecule. This diagram does not give a true impression of the relative sizes of most enzymes to their substrates. Fig. 3.3 gives this.

almost perfectly much like two jigsaw pieces fitting accurately together. This gives specificity to enzymes; each will bind only to its own substrate molecule. The substrate molecule collides with the patch and sticks to it because the catalytic site has evolved a structure so that it automatically forms weak chemical bonds with its substrate.

Figure 3.3 shows the molecular structure of an enzyme before (a) and after (b) binding to its substrate, determined by X-ray crystallography. It gives a more

(a)

(b)

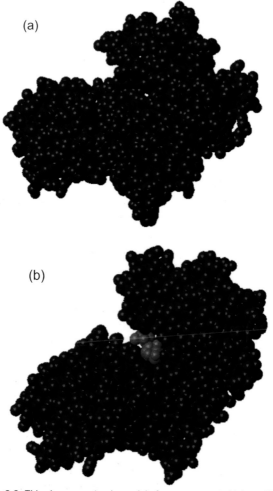

Figure 3.3: This shows an atomic model of an enzyme combining with its substrate. Each sphere represents an atom. The enzyme (hexokinase) occurs in almost all cells and its substrate is glucose, seen as the small red structure in the active site. The structure of the enzyme was determined by X-ray studies.

realistic illustration of the size relationship between an enzyme and its substrate. The enzyme (called hexokinase if anyone wants to identify it) carries out the first chemical reaction needed to break down glucose. It is present in probably every cell in your body. The glucose molecule is the small red coloured structure at the base of the opening to the catalytic site. Enzymes are usually vast in size compared with their substrates.

The rate of any chemical reaction is critically dependent on the rate of formation of the transition state of the reactant, and since the transition state is at a higher energy level than the substrate itself, activation energy must be supplied. Without an enzyme the activation energy is too high to be supplied at moderate temperatures, so the reaction cannot proceed. The enzyme cannot supply the energy, so how does it lower this barrier? The secret is in the way the substrate binds to the enzyme. As mentioned, each enzyme has a patch on its surface, the active site, which is designed to bind to its substrate rather like two pieces of a jigsaw puzzle fitting together. But there is something else about the site. It is structured so that it binds much more tightly to the transition state than to the substrate itself. This tends to force the substrate molecule towards the structure of its transition state. The net effect is that much less activation energy is needed to form the transition state; the amount is small enough for it to be supplied by ordinary heat energy present in all solutions. The enzyme therefore causes the reaction to occur (Fig. 3.4).

The mechanism is very efficient; enzymes can work at enormous speeds. One enzyme (urease) increases the rate of the non-enzymic reaction billions of times. An enzyme in your blood cells (carbonic anhydrase), which reversibly joins carbon dioxide (CO_2) and water together in the red blood cells, increases the non-enzymic rate ten million times. A single molecule of this enzyme catalyses the reaction at the rate of one million molecules of CO_2 per second.

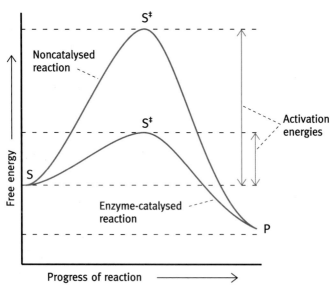

Figure 3.4: This figure shows how a chemical molecule must be raised to the top of an energy hump before it can react. A molecule so raised is called the transition state and is equivalent to a rock at the edge of cliff in Fig. 3.2. The important point is that the energy hump for a non-enzymic reaction is too high for reactions to occur under the gentle conditions of the body. An enzyme lowers the amount of energy needed for this so the reaction can proceed at body temperature. The upper dashed line represents the activation energy needed without an enzyme while the second dashed line from the top shows the much lower amount needed when the enzyme is present.

An enzyme usually brings about one reaction and one only. Cells have thousands of different reactions going on so there have to be thousands of different enzymes, each with its active site designed to bind to one substrate. There are many copies of each of the different enzymes. This adds up to a lot of protein, so a high proportion of living matter consists of enzymes. It is mind-boggling to think that evolution has produced thousands of different enzyme sites to bind with exquisite precision to the thousands of different transition states. It seems an impossible job but then we must reflect that evolution has taken millions or billions of years to achieve it. It is not surprising that evolution is a slow business.

Enzymes often need a vitamin derivative or a trace metal to function

To maintain health, several metals and vitamins are needed. Vitamins in general are usually required in milligram amounts, iron also in milligram amounts per day, and trace metals such as zinc, cobalt and copper in microgram amounts. (A microgram is one millionth of a gram). How can such small amounts have such a big effect on a body as large as a human? (Box 3.2).

Enzymes are synthesised as protein molecules in cells and in many cases the protein has to combine with another molecule called a cofactor before it can catalyse reactions. Most often the cofactor is made from a vitamin. It takes part in the chemistry of the reaction as part of the catalytic site. If your body is short of a

Box 3.2: Versatility through vitamins

In the words of American science writer Carl Zimmer:

Vitamins expand our chemical versatility. A vitamin cooperates with proteins to help them carry out reactions they couldn't manage on their own.

Here are some of the vitamins that we humans rely on to expand our chemical versatility.

Vitamin	Function	Source
A	Precursor for photo-pigments in the eye	Dietary
B1	Cofactor for glucose oxidation	Dietary
B2	Precursor for electron carriers in glucose metabolism	Dietary
B3	Component of coenzyme for electron transfer in glucose metabolism	Dietary
B5	Component of coenzyme A needed for glucose oxidation	Dietary
B6	Required for amino acid metabolism and releasing stored glucose	Dietary
B7	Required for carboxylation reactions in glucose metabolism	Dietary
B9	Precursor for coenzyme active in nucleotide metabolism	Dietary
B12	Precursor for coenzyme active in carbon transfer reactions	Dietary
C	Required for hydroxylation reactions to form collagen	Dietary
D	Converted to calcitriol, activates genes for calcium absorption	Produced in skin
E	Acts as an antioxidant, mopping up free radicals	Dietary
K	Cofactor for enzyme required for blood clotting	Dietary + gut bacteria

particular vitamin needed to activate an enzyme, then the reaction that enzyme is responsible for may be absent or may not occur at an adequate rate, and this may appear as a disease (Box 3.3).

Box 3.3: Enzymes are critical for good health

Normal bodily functions rely on thousands and thousands of enzymes to accelerate critical chemical reactions. Enzymes are proteins that are made inside the cell following instructions encoded by genes in our DNA (Chapter 5). If the instructions are faulty, the production of that enzyme can be compromised to the extent that it does not fulfil the regular level of function. For enzymes that are crucial for health, this can show as a condition or disease for which a specific array of symptoms reflects the low levels or absence of that particular enzyme. The table below outlines the function of five enzymes in the human body, and describes the condition and associated symptoms and signs that result when a genetic defect in that enzyme occurs.

Name of enzyme	Role in body	Condition	Clinical symptoms and signs
Phenylalanine hydroxylase	An enzyme found mainly in the liver which accelerates the breakdown of a particular dietary amino acid, phenylalanine	Phenylke-tonuria	Mental retardation and seizures due to accumulation of phenylalanine and formation of the toxic byproduct phenylpyruvate
Aldehyde dehydro-genase	Liver enzyme involved in breaking down acetaldehyde, a byproduct of alcohol metabolism	Alcohol flush reaction	Flushing of face and other body parts due to accumulation of non-metabolised acetaldehyde in blood and tissues

Box 3.3: (Continued)

Galactose-1-phosphate uridylyl-transferase	One of the key enzymes required to convert ingested galactose (a sugar found in milk) to glucose	Galacto-saemia	Symptoms in newborns after milk ingestion resulting from accumulation of galactose and metabolites: jaundice, enlarged liver/spleen, food intolerance and cataract development
Calcium-transporting ATPase type 2C member 1	An enzymatic pump which transports calcium ions. Calcium is needed for skin cells to form strong connections with their neighbouring cells	Hailey-Hailey disease	Blistering skin condition due to poor connections between skin cells
Holocar-boxylase synthetase	Links biotin (a B vitamin) to a variety of enzymes required for normal metabolism of dietary proteins, fats, carbohydrates	Multiple carbox-ylase deficiency	Difficulty feeding, breathing problems, skin rash, hair loss, lethargy, delayed development, seizure and possible coma due to nutrient deficiencies and disrupted cellular functions

A most dramatic example of this was in the Department of Biochemistry at Oxford University many years ago. Professor Rudolph Peters was its head, and Bill was a junior member of this department. Peters was noted for his distinguished work on how vitamin B1 (thiamine) works. It is needed as part of a cofactor for an enzyme involved in the oxidation of glucose. If thiamine is not available the enzyme cannot catalyse the essential reaction and glucose therefore cannot be oxidised normally to produce ATP. The brain gets its energy mostly from glucose and is very sensitive to thiamine deprivation. Peters once showed Bill the spectacular effect of thiamine. He had induced thiamine deficiency in pigeons by feeding them only on polished rice, which lacks the vitamin. A pigeon lay spread-eagled on the floor of its open cage, unable to move or stand. After a small injection of thiamine, in a matter one or two minutes it blinked, stood up, and flew vigorously out of the cage around the laboratory, with Peters beaming happily at it. His assistant looked on also, wondering, Bill suspected, how he was going to catch it. In humans, vitamin B1 deficiency produces the condition known as beriberi.

In other cases, some enzymes may require a metal atom such as copper, cobalt or zinc to be attached to the enzyme before it is active. Most enzymes involved with ATP as a substrate require magnesium. Several trace metals are required in minute quantities for normal health. An enzyme that requires a metal atom as part of its active centre cannot be fully active if the metal in question is deficient in the body, again possibly causing a disease, and if the metal is not available it is as if the enzyme is not there.

It is easy to see why such small amounts of vitamins and metals have such a big effect on human health. A single molecule of a vitamin or an atom of metal can activate a single molecule of a particular enzyme, and that enzyme molecule can now catalytically bring about reactions in huge numbers of substrate molecules – often at rates of hundreds of thousands per second. So very small amounts can result in a large number of essential chemical reactions in the body.

Most vitamins act as enzyme cofactors, but there are a few such as vitamins C, D and E that have other roles. Vitamin C is an antioxidant (see page 122 and Box 3.2); vitamin D is really a hormone or chemical messenger controlling the absorption of calcium, though it reportedly has several effects on health. Vitamin E is also an antioxidant but is fat-soluble, whereas vitamin C is water-soluble.

If the activities of enzymes were not regulated there would be metabolic chaos

The picture so far painted of a living cell is of a mixture of thousands of chemical reactions going on at the same time; the reactions are controlled to give an organised chemical machine whose activities match the ever-changing needs of the organism. If you are playing squash, much more energy is needed than if you are asleep – as much as five or six times more. To generate the requisite amount of ATP for vigorous work, the rate of food breakdown, or oxidation, increases. You get hot when you exercise because, in addition to producing ATP, these reactions also release heat. When you rest, the process is slowed down. In the metabolic pathways generating ATP – that is, the sequences of enzyme actions that oxidise food – certain enzymes control the rate of the oxidation. As well as binding to their substrates, these enzymes have additional sites or patches on their surface that bind to molecules quite different from their substrate. These are known as regulatory sites. Some of the enzymes in the food oxidation pathway bind ATP on their regulatory sites. Nothing happens to this bound ATP but it causes the enzyme to change its shape slightly, and this slows down or stops its catalytic activity. The rationale is that if there is plenty of ATP around there is no need to produce more, so the whole pathway is slowed down by feedback control. As the ATP is used up and its level falls, the ATP molecule leaves the regulatory site and the oxidation-producing ATP process speeds up again. Different metabolic pathways in the cell communicate with each other in this way to regulate the entire chemical activities of the cell, ensuring that there are no shortages and no pile-ups of products. It might be likened to a car factory – if more cars are produced than are sold, the assembly lines are shut down until more are needed, and at the same time the delivery of components from other factories is likewise reduced. Although enzymes are lifeless molecules, they work in such a sophisticated way that they almost give the impression of being intelligent molecular machines which automatically adjust to the changing needs of the organism.

There is an additional metabolic control system, regulated by hormones. Suppose you are in a state of alarm, such as if you were being chased by a tiger. Your brain signals the adrenal glands (near your kidneys) to flood the body with the hormone adrenalin. This rings the alarm bells in your cells and the normal controls are over-ridden to prepare the body for emergency action. ATP production is maximised. The hormone activates enzymes that cause the liver to pour out glucose into the bloodstream to supply your muscles with all the fuel they might need for a violent response, such as running away. The muscles themselves break down glycogen (a storage form of glucose) to ensure there is adequate fuel to oxidise, and the system oxidising it is activated to maximum speed. This sequence is called the fight or flight reaction. Everything is done to make sure that there is a maximum supply of ATP to increase your chance of escaping from the tiger. Unfortunately, from your point of view, the tiger's body has the same response to maximise its chances of catching you. On the whole therefore it is best not to place too much reliance on your protective biochemical response and, if possible, not get into the situation in the first place.

Some poisons inhibit the activity of an essential enzyme. Cyanide, for example, is so deadly because it combines with the final enzyme in the sequence of reactions that oxidises foodstuffs to carbon dioxide (CO_2) and water. It has the same effect as deprivation of oxygen in blocking ATP production.

Many inherited (genetic) diseases can be explained by the absence of a specific enzyme. The production of an enzyme depends on instructions coded in our DNA by our genes (see Chapter 5). DNA and the genes it carries have to be copied for reproduction to occur, and occasionally the copying is not perfect and the corresponding enzyme is faulty. If that enzyme is crucial for health then its absence may cause a disease.

Why don't digestive enzymes digest the body?

The food that we eat is largely composed of macromolecules such as starch and protein. Macromolecules are assembled in organisms by joining together huge numbers of small 'building block' molecules. Starch is made from glucose; proteins from amino acids. Sucrose, for example, a main dietary component, is made of two sugars (glucose and fructose) joined together. In general the intestine can only absorb small molecules. The job of digestive enzymes is to break the macromolecules or things like sucrose into their component parts, which can be absorbed into the circulation in the body. Fat is also broken down to smaller components; to enable enzymes to get at the insoluble fat, bile salts, a form of detergent, make the fat molecules more available to attack. The digestion of the foodstuffs involves reactions that are energetically downhill so they go to completion.

In addition to enzymes breaking down the main food constituents mentioned, other enzymes exist in the intestine to break down most other molecules found in food, such as DNA. The remarkable aspect of digestion is that it breaks down food that is virtually the same as the materials your own body. The enzymes secreted into the intestine have to be made in your own cells, for example those of your pancreas. Why don't they digest your own body? There are precautions against it. In the pancreas the digestive enzymes are made in separate compartments of the cells in an inactive form. They don't become active enzymes until they reach the intestine.

The intestine itself is protected against digestion by a layer of mucus which prevents the enzymes attacking it.

Enzymes are one major class of proteins, but there are others. In a real sense proteins are the essence of life. The next chapter is to tell you all about them.

4

Proteins: the wonder molecules of life

It is difficult to think of anything in the mechanism of life that does not depend on proteins. All the chemical reactions of life are brought about by enzymes, the special class of proteins discussed in the previous chapter. (A few rare exceptions to this will be mentioned later in Chapter 5.) There are many thousands of different proteins in a human, probably between 250 000 and a million as a rough estimate, each with its own unique structure, the blueprint for which is stored in coded form in a gene.

The list of protein functions is very long. Apart from enzymes, there are contractile proteins responsible for muscle action and transport proteins such as the haemoglobin in blood cells that carries oxygen from the lungs to tissues. There are motor proteins, which pull molecular loads around the cell, running on protein tracks. Their functions are to transport molecules made in one part of the cell to distant parts of the cell that need them. As in cities, there are several different makes of molecular motors differing in the loads they carry and how they move. Fireflies have a protein which, using ATP, makes flashes of light to attract mates. There are structural proteins like collagen, a major component of tendons, which are very strong, and the very tough cartilage pads that withstand the great pressures between bones at joints. Other proteins transport sodium and potassium ions across nerve membranes and allow the propagation of nerve impulses. There are protein hormones such as insulin which regulate metabolism (Box 4.1). A complex animal like a human has a complex system of chemical signalling between its cells; signals are picked up by external protein receptors acting as aerials and passed on into the interior of cells. Other proteins in the cell accept the message and respond to it. Antibodies of the immune system are proteins that protect you from infections (Box 4.2). To give extreme cases of versatility, hair is a protein and so are horses' hooves, spiders' webs, and silk.

How can one type of molecule have so many different functions?

To understand this we must look at their structure. Proteins are very large molecules compared with molecules like glucose. Protein molecules are made by joining together large numbers of small units called amino acids. A small amount of chemistry is needed here. An amino acid is so-called because it contains as part of its structure an amino group (NH_2 – a nitrogen atom with two hydrogen atoms attached) and an acid group (COOH – a carbon atom with a double covalent bond to an oxygen atom and with a single covalent bond to an oxygen atom that in turn is bonded to a hydrogen atom). There are many possible amino acids but the same 20 different ones are used as the units from which all proteins are made; they are what Francis Crick, the co-discoverer of DNA structure, called the 'magic 20'. The amino acids in the magic 20 all have identical short chains with the structure H_2N-CH-COOH in which the amino group, the acid group and a hydrogen atom are all attached to the same carbon atom.

Box 4.1: Protein drugs to improve sporting performance

Thanks to modern science and medicine, the structures and functions of many bodily proteins are well understood. This knowledge can be used to treat individuals with diseases in which a particular protein is missing or defective. For example, people with type 1 diabetes – in which the beta cells of the pancreas fail to secrete the protein insulin to maintain blood sugar levels – can inject themselves with manufactured insulin as a compensatory drug. Even though they are medically healthy, some sports people also administer

Box 4.1: (Continued)

proteins to boost biological activities in their bodies. Sadly, many do this even if there have been no studies in humans to prove efficacy or determine safe levels of administration.

Four proteins – growth hormone, erythropoietin (EPO), insulin-like growth factor (IGF)-1 and insulin – and their use to boost sporting performance – are shown in the diagram below. A summary of natural production sites, actions in the body, perceived or known benefit to athletes, source when used as a drug, and a comment relating to testing is presented for each of the proteins.

Growth hormone
- Produced by pituitary gland
- Increases protein production
- Improves muscle bulk
- Strengthens bones and tendons
- Manufactured or collected from dead bodies for use as a drug
- Used since 1980s to improve athletic performance
- Notoriously difficult to detect abuse

Insulin-like growth factor (IGF)-1
- Produced mainly in liver
- Mediates some of the actions of growth hormone
- Increases protein production
- Improves muscle bulk
- Manufactured for use as a drug
- Used by professional athletes and baseball players
- No urine test, can only be detected using blood test

EPO
- Produced mainly in the kidney
- Increases red blood cell production
- Enhances oxygen-carrying capacity
- Manufactured for use as a drug
- Used by competitors at the Tour de France
- Urine test used to detect EPO abuse by Lance Armstrong

Insulin
- Produced by the pancreas
- Promotes uptake of blood glucose for use or storage in cells
- Inhibits protein breakdown
- Improves muscle bulk
- Manufactured for use as a drug
- Relied on to control diabetes
- Some use in bodybuilding
- Difficult to detect abuse due to very rapid clearance when used as a drug

Box 4.2: Antibodies are protein recognition molecules

One of the examples of molecule recognition you may have already heard of is that involving antibodies. Antibodies are a class of proteins that bind very specifically to other molecules. The body's production of antibodies provides a mechanism through which the immune system can detect and mount a response against proteins that are foreign and may present danger.

The process of immunisation relies on the guided production of antibodies for disease prevention. The diagram below provides a simplified representation of what happens when you are immunised against human papilloma virus (HPV).

1. The vaccine contains small non-infectious protein fragments of the virus and these are injected into the muscle.
2. A pre-existing B-cell binds to a specific virus protein through an antibody-like cell receptor and the cell becomes activated.
3. The activated B-cell multiplies and each daughter cell releases thousands and thousands of free molecules that are also shaped to recognise that particular virus protein. These molecules are called antibodies.
4. If you later become exposed to the virus through natural infection, the antibodies that are already in existence bind to the viral protein that matches the protein previously encountered in the vaccine.

Box 4.2: (Continued)

5. The protein with antibody bound to it becomes a target for destruction by other elements of the immune system.
6. Other proteins in the body are not targeted for immune destruction, as their three-dimensional conformations do not bind with the antibodies.

Professor Ian Frazer was awarded Australian of the Year in 2006 for his work linking HPV infection with cervical cancer and development of a vaccine against HPV.

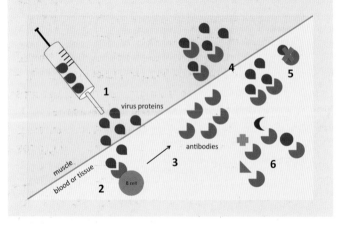

Also attached to the central carbon atom is a small chemical structure or side chain designated as R, for radical (Fig. 4.1). Different amino acids have different radicals; they differ in size and shape and in electrical charge.

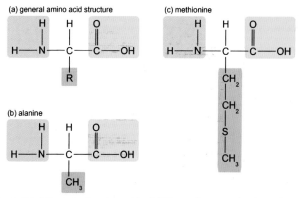

Figure 4.1: Structure of amino acids. (a) The amino group is shown in green, the acid group in yellow and the side chain – the part which differs for each individual amino acid – in pink. (b) and (c) Examples of two different amino acids.

Basically proteins are simple structures consisting of one amino acid after another being linked together via chemical reactions between their amino and acid groups to form the protein or peptide chain known as the 'backbone', with the different side chains (represented as R) projecting from it. (When the amino and acid groups react to join two amino acids together, they form the chemical link -CO-NH- and a molecule of water (H_2O) is extracted.) A section of a protein (polypeptide) backbone is shown below, with the bonds between different amino acids in red.

The length of the peptide structure depends on the protein involved. Insulin is one of the smaller proteins with 51 amino acids, but different proteins, or polypeptides, have different numbers of amino acids linked together. Most proteins have several hundred, but some range up to a few thousand.

With a few exceptions, all of the magic 20 are present in each protein. Each protein has its own particular mixture of the different amino acids and its own particular order in which they are arranged in the protein chain. This is known as the amino acid sequence of the protein. The particular amino acid sequence of each protein is specified in coded form by the gene for that protein. It is this that is the stored wisdom of billions of years of evolution. One of the vital points in the whole system of life is that there is virtually no limit to the number of different protein structures that in principle could exist. With variation of chain lengths, amino acid composition and amino acid sequence, astronomical numbers of different protein structures are theoretically possible. This is one of the reasons for protein versatility. During reproduction of organisms, errors, called gene mutations, can cause a protein to have a faulty amino acid sequence so that vast numbers of variations are created over long time spans. Evolution is primarily based on these random variations in protein sequences, with natural selection picking out the beneficial ones for preservation and eliminating the bad ones. Beneficial ones increase the chance of an organism reproducing itself; bad ones do the opposite.

It is useful to imagine the 20 amino acids as the first 20 letters of the alphabet. You can think of the many proteins as thousands of different words, each being hundreds or thousands of letters long. A system for writing down protein sequences has been developed in which each amino acid is represented by a capital letter; A for the amino acid alanine and M for for the amino acid methionine, and so on. Protein structures can then be represented simply as a string of letters (in the case of insulin by 51 letters).

A single wrong amino acid out of hundreds can cause a genetic disease

In each protein the amino acids must be arranged in the correct sequence. Each protein must be 'spelled' correctly, to return to the word analogy. Suppose a protein is made with a single mistake in it – a single amino acid in the sequence being incorrect. In some cases it may have no effect; or it may cause the protein to improve its function or even result in it having a new and beneficial function. If this improves the chances of the organism having a better chance to survive to reproductive age, the mutated gene will be passed on to offspring. Natural selection will select the changed gene. Or it may prevent the protein from functioning, in which case it may be lethal and the offspring is never born. In between the two extremes it may cause a disease in the person with the defective protein (see Box 3.3). If either of these events reduces the chance of the organism reaching reproductive age, the 'bad' gene will tend to be eliminated by natural selection.

A classical example is the genetic disease called sickle cell anaemia caused by a single incorrect amino acid in haemoglobin, the protein in your red blood cells that carries oxygen. The mistake causes the protein to crystallise as long rods, which distort the red blood cell into a sickle shape. The abnormal cells tend to block capillaries in the blood circulation and also to break up so that there are not enough red blood cells circulating (anaemia). The disease, to use Nobel Prize-winning biochemist Linus Pauling's comment, is due to a sick molecule. Oddly enough, to illustrate the complexity of biological interactions, in certain parts of the world this mutation is beneficial to survival and sickle cell disease becomes prevalent due to its genetic selection. This occurs where there is a very lethal form of malaria. Sickle-shaped cells tend to prevent the malaria developing. In people with only one copy of the sickle gene and one normal copy, there is enough normal protein so that they do not suffer from anaemia, but enough sickle cell protein to have a protective effect against malaria. In this way, the sickle cell gene is beneficial so far as survival goes.

There are many genetic diseases arising out of mistakes in the amino acid sequences of proteins and some of these are distressing, such as cystic fibrosis. Unfortunate as these are, mistakes in protein synthesis are the feedstock for evolution. Such mistakes give a supply of new protein structures for evolution to choose from. Natural selection over long time spans is infallibly efficient in preserving what is good in random variations and eliminating deleterious mistakes. But it is utterly ruthless and is concerned only with the survival of the fittest.

Darwin in his book *The Origin of Species* summarised the situation as follows: 'It may metaphorically be said that natural selection is daily and hourly scrutinising, throughout the world, the slightest variations; rejecting those that are bad and adding up all that are good, silently and insensibly working, whatever and whenever opportunity offers, at the improvement of each organic being in relation to its organic and inorganic conditions of life.' The Nobel laureate François Jacob put it more succinctly when he said, 'Evolution is a tinkerer'.

Folding of the proteins

A human being develops from a fertilised egg into a baby. This development depends on the assembly of proteins produced. As we have seen, proteins are made in living cells initially as microscopically thin polypeptide threads somewhat like a spider's web (which itself is a protein). It does not seem possible for such molecules to assemble into a solid three-dimensional structure such as a human.

It is comparatively easy to see how it is possible to give instructions for making specific linear sequences of amino acids. In principle it is not basically different from sending a linear message by Morse code in which dots and dashes are used to indicate letters. Somehow the thin protein threads, as formed in the cell, have to be converted into the three-dimensional structure of the human body. The development of the embryo occurs by

an automatic process. Once the correct proteins are made
in the correct order and amount, the correct development
occurs because the chemistry of the polypeptide chains is
such that assembly of the organism follows.

Most proteins in the body are compact globular
structures, quite unlike their initial thread-like nature.
Yet the only known structural information in the genes is
the amino acid sequences of proteins. The answer to this
paradox is that the polypeptide chains, as soon as they
form, spontaneously fold up to become three-
dimensional, compact proteins. Fig. 4.2 shows a

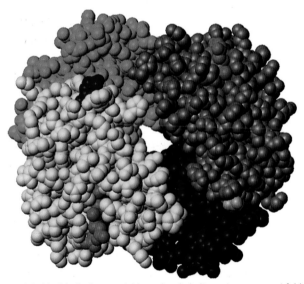

Figure 4.2: Model of a haemoglobin molecule in its mature compact folded
form. Haemoglobin is the protein in your red blood cells that transports
oxygen from the lungs to the tissues. It has four subunits, shown in green,
yellow, light blue and dark blue. The small patches in red are the haem
molecules that bind oxygen.

molecular model of the protein in your red blood cells called haemoglobin, in its final, compact form. Such space-filling models show the actual structures of molecules as determined by a technique known as X-ray diffraction. The haemoglobin chain takes up a helical (spiral) configuration. It is not a random folding; the function of each protein is dependent on the folds being exactly correct in its three dimensions, and each protein is different.

A folded protein is held in its mature form by chemical bonds between different sections of the chain; these stabilise it. It is the amino acid sequence that determines which bonds form and therefore determines the folding. The bonds involved are *not* the covalent bonds described in Chapter 1 (page 23) but are of a different type called *weak* or *non-covalent* bonds. The term 'weak' might imply that they are not very important. In fact they are yet another of the major 'secrets of life'. DNA function and the entire genetic system, protein formation and just about all aspects of life depend on them. We need to have a small diversion here to explain what they are.

The most important type of chemical bond in the present context is the hydrogen bond. Its nature can be best illustrated by looking at the water molecule. This has two hydrogen atoms joined to the oxygen atom by covalent bonds; the two bonds are at an angle, giving a triangular shape. The oxygen atom is greedy for electrons and attracts them more strongly than does a hydrogen atom. The effect is that although the pair of electrons forming each covalent bond is shared between the

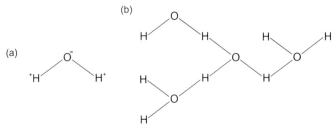

Figure 4.3: The structure of water. (a) A single water molecule showing that the oxygen atom attracts more than its share of the electrons that form the covalent bonds between the atoms. This gives the hydrogens a partial positive charge and the oxygen a partial negative charge. (b) In bulk water the molecules are linked together by the weak charges, giving water a structure. This has a profound effect. Without it the world would be a very different place so far as life is concerned.

hydrogen and oxygen atoms, the oxygen attracts more than its share, giving it a partial negative charge (since electrons carry negative charges). By the same token, the hydrogen atoms have a partial positive charge. In bulk water the negatively charged oxygen atoms are attracted to the hydrogen atoms of adjacent molecules, forming a weak bond as illustrated in Fig. 4.3.

This happens between the trillions of water molecules, causing the whole water mass to stick together in a fragile network. Such bonds are so weak that, unlike covalent bonds, they spontaneously form very easily and they break easily; because of their weakness very little heat is needed to disrupt them. They are constantly forming and breaking all the time, giving a flickering stability, but there are always sufficient numbers at any one time to give a weak coherence to the body of water.

Without this hydrogen bonding of water, the world would be a very different place and possibly unsuitable for life. To give a little everyday normality to the subject, the hydrogen bonding of water greatly raises the boiling point of water above what we could expect from the size of the molecule; the hydrogen bonds make it more difficult for molecules to burst free from solution into the air. This enables you to make good tea, but much more importantly it enables water to flow upwards in tall trees from the roots to the leaves via the plants' thin water transport tubes or xylem. When a water molecule evaporates from the leaf, as it enters the air it exerts an upward pull via hydrogen bonding on the molecules of the water in the xylem.

To return to protein folding, each amino acid in a protein chain is capable of forming hydrogen bonds with other amino acids in the chain. It is necessary to form these for the folding to have the greatest stability. The protein chain spontaneously adopts a folded structure to achieve this. The reality of protein folding and its nature is reflected in a familiar event – frying an egg – which most of us have done. Egg white is largely made of the protein albumin. These are neatly folded compact globular molecules soluble in water; egg white is transparent. As heat is applied to the frying pan the egg white starts to become cloudy, and quickly turns into an insoluble white mass. The mild heat has no effect on the covalent bonds of the protein chain holding the amino acids together, which remain intact, but the weak non-covalent bonds responsible for the folding are torn apart

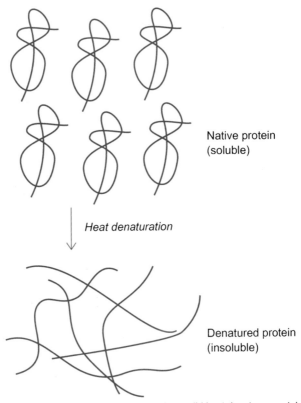

Native protein
(soluble)

Heat denaturation

Denatured protein
(insoluble)

Figure 4.4: This is a diagram illustrating how mild heat denatures proteins such as happens to egg albumin in the frying pan. The weak bonds holding the molecules in a folded form are disrupted and the protein chains become entangled into an insoluble mass. Note that the folded shape of the molecules in this diagram is hypothetical.

by the vibrations induced by the heat. The folded structure is destroyed and the unravelled protein chains become irreversibly entangled with one another, forming the insoluble white mass of a fried egg as illustrated diagrammatically in Fig. 4.4. You can imagine each albumin molecule to be a length of string neatly wound into a small ball in a specific way. The balls remain separate from one another in the normal albumin protein. The vibrations caused by mild heat unravel the winding of the string of each molecule and they become entangled together.

Complex structures of life form from protein molecules by a process of self-assembly

One of the most remarkable things about complex organisms such as a human is how the developing embryo forms the complex organs of the organism. The finished folded proteins self-assemble into the structures of the living cells. As already stated, if the correct proteins are made in the right quantities at the right time, then the living structures assemble automatically. If a developing fertilised egg produces rabbit proteins a rabbit will result. Whale proteins produce a whale. The simplicity of this concept in solving one of the most difficult problems in life is satisfying.

What causes proteins to assemble correctly? A folded native protein has an individual shape with bumps and valleys on its surface that make it recognisable in fine detail. Proteins can 'recognise' other proteins. Just about everything in life depends on this. By 'recognise' we mean that proteins can combine with other specific molecules – each recognises its cognate molecule, the one that the protein evolved to combine with. Most proteins have on their surfaces one or more special small patches.

Proteins destined to combine with each other have complementary patches that automatically bind together by weak non-covalent bonds. The fit between the molecules must be close because weak bonds form only if the patches on the pair of proteins are close together. For a protein to combine with its cognate molecule they must have close matching of shapes with correct types of chemical groups able to form bonds at several contact points. A single bond is not sufficient; there must be several. These requirements mean that protein–protein associations are specific – the chance of the protein recognising an incorrect molecule is remote, so that only appropriate bindings occur.

This recognition by protein molecules of other molecules is important in almost all aspects of life. Gene function, the immune system protecting us from bacterial and viral invasions, nerve action, senses of taste and smell, muscular movement, enzyme activity, hormone action and much else, all depend on specific recognition of other molecules by proteins (Box 4.2).

The evolutionary record is written in the amino acid sequence of proteins

The history of the evolution of living organisms from primitive forms of life to more advanced ones was deduced originally from studies on fossils and comparative anatomical studies. Evolutionary trees have been constructed showing how different evolutionary lines have diverged from common ancestors. There is a wealth of evidence showing that the evolutionary history of organisms is also reflected in the amino acid sequences of proteins. For example a protein involved in DNA function, present in all forms of life, has almost exactly the same sequence in the corresponding protein found in organisms as evolutionarily distant as peas and cows, suggesting that they arose in both from a common ancestor. In the haemoglobin protein that carries oxygen in the blood, out of 153 amino acids in the protein, the sequences in human and chimpanzee are different in only one amino acid. More distant relatives have more numerous differences. By studying the amino acid sequences from individual proteins in many organisms it has been possible to construct evolutionary trees which largely agree with those based on more conventional methods. DNA sequences give similar results, which is not surprising given that the amino acid sequences are determined by the DNA of genes.

From all of this it is clear that the central role of DNA in life is to instruct cells on the correct amino acid sequence of each protein. The next big question – how it does this – is discussed in the next chapters.

5

The genetic problem and DNA

Billions of years of evolution, based on gene mutations and natural selection of the fittest, has produced many different life forms as diverse as bacteria and humans. The wisdom obtained by all this experimentation is encoded in genes so that the information can be passed on from parents to offspring. Your antecedents go back billions of years. This handing on of information has had to be done over these vast time spans; the continuation of life depends on it.

The handing down of this information to the next generation is done by the same basic mechanism in all forms of life from bacteria to humans. There are of course differences in the details of how it is done. The sex life of a bacterial cell or that of a plant is different from that of a human, but the differences are at a mechanical level. At the molecular level it always involves DNA. (Several viruses use RNA but its principle and structure is basically the same as DNA – see page 99.)

To understand how DNA plays its central role in life we must look at its structure. It exists in the cell as very long molecules called chromosomes. In humans there are two sets of 23 pairs of these, 46 chromosomes in all – one set from each of the two parents. A DNA molecule is a very long fine thread. A gene is a short section of such a thread. Each cell in your body has in its nucleus a complete set of genes (mature red blood cells and gametes (sperm and eggs) excepted). The total length of DNA per human cell amounts to approximately 2 m packed into a microscopically small nucleus. Since you have trillions of cells, the total length of DNA in your body is vast.

During development of an organism from a fertilised egg, the embryo grows in size by cell multiplication. Division of a cell produces two daughter cells that grow to full size, and these in turn divide. Cell division on a large scale occurs in the body. It occurs in organs of the body to replace cells that have died; red blood cells are replaced every 100–120 days in humans and turnover of lining the intestine occurs at a high rate. Before cell division occurs, the total amount of DNA in a cell, and therefore the number of chromosomes, is doubled by copying, so that each daughter cell inherits a full complement of genes.

DNA is a nucleic acid, the term deriving from its location inside the cell nucleus. As is the case with all macromolecules in life, DNA is made by linking together large numbers of small units. The small units from which it is made are called nucleotides, and all have the structure:

P–Sugar–Base

P here is phosphate; the sugar is deoxyribose, which is quite similar to glucose (Fig. 1.3) but has only five carbon atoms and lacks one oxygen – hence the deoxy term. The term 'base' is simply a collective term for a type of molecule whose structure will be shown later. Putting all this together DNA is deoxyribonucleic acid. It consists of vast numbers of these nucleotides linked together. The DNA in each of your cells contains 6.4 billion of them. There are only four different bases in DNA, usually abbreviated to their initial letters as A, T, G and C, but just for curiosity their full names are adenine, thymine, guanine and cytosine.

In DNA the P–Sugar parts are linked together to form the 'backbone' from which the bases project from the side, as shown in Fig. 5.1. Despite its awesome role in life, the DNA chain is chemically a somewhat boring structure – a long sequence of alternating phosphate and sugar molecules – millions and millions of them making up the backbone chain. The backbone, which has no variation throughout its entire length, cannot have any information encoded into it, so the real interest of DNA lies in the bases attached to the sugar group of each nucleotide unit; these stick out to the side of the chain. The information or blueprint of a living organism resides in the sequence of the four different bases along the chain. This contains the information in the DNA which causes the correct development of the offspring, which, when you come to think about something as complex as a human being, is a pretty impressive lot of information, and all done by the arrangement of four different bases along a very long molecule.

DNA in the cell does not exist as a single strand, however, but as a double strand; two backbones are attached side by side via hydrogen bonding of the bases. (Hydrogen bonds are described in Chapter 4, page 89.)

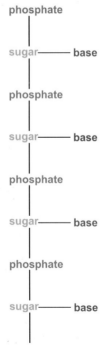

Figure 5.1: A single strand of DNA.

The structure resembles a ladder, the two backbones representing the uprights and each rung made of a pair of bases linked together (Fig. 5.2a). However, as shown in Fig. 5.2b, the uprights of the ladder are twisted around each other to form the famous double helix with the rungs (base pairs) still *in situ*. Figure 5.3 shows a space-filling model of a section of DNA; don't try to understand it unless you want to – it is here just to show what a beautiful molecule DNA is. You probably can make out how the two chains wind around each other in the double helix. The grooves in the structure allow gene control proteins in the cells to access (read) the bases,

(a)   (b)

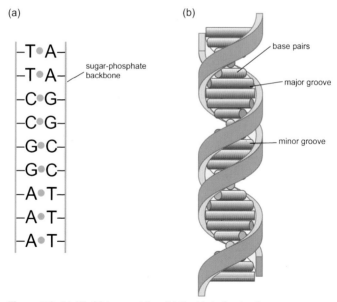

Figure 5.2: (a) AT–GC base pairing. (b) Double helix structure. (A = adenine, T = thymine, G = guanine, T = cytosine).

which is the first step in using the information content of genes.

The structure of DNA was discovered by a pair of research workers in the famous Cavendish Laboratories in Cambridge, England. James Watson was a young American cell biologist who came to England as part of his postdoctoral studies. Francis Crick was a physicist who had spent the war years in naval research on mines to blow up enemy ships. Crick was unsure what to do in his future career but finally also went to the Cavendish Laboratories to do a PhD. The details of their partnership on DNA structure and their subsequent award of a Nobel Prize are well known.

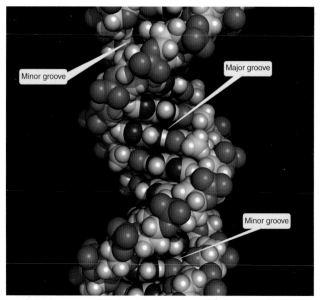

Figure 5.3: A more realistic model of a section of double-stranded DNA.

Replication of DNA

The concept involved in this process is perhaps the most important of all in the business of life.

As mentioned, DNA contains only four different bases in its structure – A, T, G and C. Erwin Chargaff in New York had discovered that in all DNAs from different organisms, the amount of A always equalled the amount of T and the amount of G always equalled the amount of C. However the total amount of A + T and of G + C varied in DNA from different organisms. The two strands of DNA are held together by hydrogen bonds between the bases on different strands. But the crucial part of the story is that these bonds are formed only between A and T and between G and C. In DNA other pairings do not exist; A is not paired with G and C is not paired with C. The A–T pair is held together by two hydrogen bonds, the G–C pair by three of them. The A–T bases have complementary shapes that fit together, allowing hydrogen bonds to form between them, and similarly for the G–C pair as shown in Fig. 5.4. This specific pairing is known as Watson–Crick base pairing. The hydrogen bonds are shown as broken lines. Full structures are shown of the two base pairs. It is reasonable to speculate that if extraterrestrial life is ever discovered something similar to A–T, G–C base pairing might be the secret of that life too. You can easily see that the molecules in each base pair are so shaped that they fit together very neatly. Just marvel at the fact that the five short broken lines representing weak hydrogen bonds between the bases make it possible for you and all life to exist. It's a bit mind-boggling put like that. I will shortly explain why this simple device of specific base pairing is so crucial to the existence of life.

When Watson and Crick deduced this base pairing, all of life became much more understandable. And it's such a simple device. Just remember that when you watch all those life forms on wildlife television programs it all depends on Watson–Crick base pairing.

Base pairing occurs spontaneously. The job of genetic material is to store information to be passed on to offspring in molecular form. Because so much information is needed in even lowly organisms, the molecules need to be very large. The other need is that the molecule can be accurately replicated so that a complete copy of the genetic material can be passed on to each offspring. This is a colossal task. As already stated

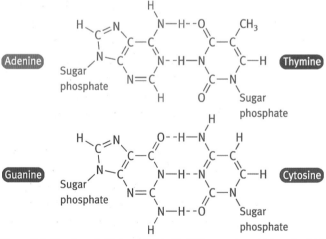

Figure 5.4: The famous Watson–Crick AT–GC base pairs, which well might be called the secret of all life on Earth.

above, in the total DNA of a human cell there are 6.4 billion bases. And yet the solution to the problem must have been developed very early after the origin of life and therefore must be simple. At that stage elaborate mechanisms could not have been developed.

The solution to the problem lies in the specific Watson–Crick base pairing. Let us consider a double-stranded piece of DNA which is to be replicated in the cell. The hydrogen bonds are strong enough to link the base pairs together but weak enough to be easily separated. They are exactly right, a phrase from the story of a man who gave his gardener a bottle of whisky. On enquiring a few days later how he liked it, the gardener replied that 'it was exactly right; if it had been any worse I couldn't have drunk it and if it had been any better you wouldn't have given it to me'. If the hydrogen bonds were any weaker they wouldn't hold the pairs together and if any stronger they couldn't be easily separated by the proteins evolved to do this. Once the strands are separated, the two single strands act as templates for producing new complementary partner strands. Bases of free nucleotides in the cell automatically pair up with the exposed bases on the template strands; wherever there is an exposed A, a T floating around in the cell automatically joins to it by hydrogen bonding. Similarly G–C base pairs automatically form. This automatically arranges that the incoming bases are lined up so that, when they are linked together, each old strand is now accompanied by a new strand, and all that is needed is to link the units together, which is done by an enzyme. It enables your cells to copy the 6.4 billions of bases in your

DNA. It explains how life is reproduced. In their first publication of their work in the science journal *Nature* about the structure of DNA, Watson and Crick added the famous comment that it had not escaped their notice that specific base pairing offered a possible copying mechanism for the genetic material. That's all they said at the time about the solution of almost the greatest problem in biological science.

After Watson and Crick made their discovery, Crick, an outgoing person, announced to the patrons of the Cambridge pub The Eagle where he had lunch that they had discovered the secret of life. People who had happened to go in for a beer were the first members of the public to hear of the momentous discovery. Crick was not exaggerating. He and Watson were the first persons to understand what has been the basis of life for over 3 billion years. Their discovery has transformed biological sciences.

It should be added that while the principle of DNA replication is simple, the enzymic apparatus that carries it out is complex. Initially the problem of copying would have been simpler because in the early primitive life forms the amount of DNA would have been much smaller and the chances of mistakes correspondingly less. As organisms grew more complex, devices were evolved to ensure that mistakes were eliminated as much as possible. In human cells the enzymes involved in linking the nucleotide units together check that the correct base has been added, and, if not, the wrong one is removed and the error corrected.

So far I have described the importance of specific base pairing in DNA replication. But it is also the basis of

Box 5.1: Engineering genes for human benefit

Left to nature alone, new genetic combinations (genotypes) can only appear as a result of random mutations in DNA, or if the rearrangement of parents' DNA during breeding introduces a new DNA sequence into offspring which was not present in the parent. Such rearrangement is a normal part of reproduction and takes place each time a new offspring is made. Unless you are an identical twin, this normal rearrangement is why you are not exactly the same as your brothers and sisters, even though you both have the same parents (you have different genotypes).

Because we humans have developed a detailed understanding of how genes work, and have many tools at our disposal to manipulate DNA, we now have the capacity to introduce new genes into cellular DNA. This approach is referred to as genetic engineering, also called genetic modification or GM.

In most GM, one or several genes that give a plant or animal desired characteristics are selected and artificially inserted into DNA early in development. The resulting plant or animal is then bred to maturity, and assessed for health and to determine the success of the procedure. Some more sophisticated forms of GM involve the transfer of genetic material to selected populations of mature cells.

Many scientists and others in our communities are excited about the possibilities of GM. For example, it has allowed scientists to create 'golden rice', in which two genes have been inserted into the DNA of normal white rice to enhance production of the important nutrient β-carotene (pro-vitamin A) beyond the early

Box 5.1: (Continued)

stages of rice maturation. The GM 'golden' rice was designed in order to counter childhood malnutrition in countries such as India, Vietnam, Bangladesh, the Philippines and Indonesia.

However some people in our communities are fearful of GM. For many, there is a sense of extreme discomfort associated with the idea of inserting chosen genes into 'natural' DNA, especially if it involves transferring genes from one species into another. Other concerns relate to lack of long-term testing of health and environmental implications, and ethical issues such as the possibility of creating 'designer babies' carrying genes to enhance the chance of desired characteristics such as high intellect or sporting prowess.

the DNA-directed synthesis of all proteins. And as stated earlier, virtually all processes in life depend on correct production of proteins. Oddly enough, specific base pairing also makes possible most of genetic engineering (Box 5.1) and techniques such as DNA fingerprinting.

6

Deciphering the genetic code

The problem I now want to talk about is another one which can seem so difficult that it could never be solved. How on earth could a cell about the size of a full stop on this page take a long DNA molecule and read the instructions encoded in a sequence of four bases and translate these into the amino acid sequence of thousands of different proteins?

Remembering that these proteins in turn produce a human being, the whole business would seem like science fiction were it not for the fact that we do exist. Let us start by looking at a gene – a small stretch of an enormously long DNA molecule. There are many genes in the human genome – about 20 000 to 25 000 – the exact number is not known. In between the genes on the chromosomes there are simply sequences of bases that are not part of a gene.

The solution to this seemingly impossible problem of how a cell can follow the instructions of the genes is in principle very simple, and once again it is based on specific Watson–Crick base pairing.

In bacteria the DNA simply lies in the cell with all the other components, but in 'higher' cells – those of animals and plants – the DNA is confined to the cell nucleus, which is surrounded by a membrane (see Fig. 2.2). In all cells, proteins are synthesised by small spheres called ribosomes which exist in large numbers. In higher cells these are outside the nucleus, quite separated from the DNA inside. But how do ribosomes receive instructions from the genes on the correct amino acid sequence of the protein for which that gene is responsible? There's a problem here because the gene cannot leave the nucleus and the ribosomes cannot enter it. So how do the two communicate? The answer is that the gene sends a message to the ribosomes; this exits the nucleus and attaches to the ribosomes outside.

What is the nature of this message? It is a copy of the gene, but this needs to be qualified. A gene is part of a double-stranded DNA molecule. Only one of the two strands, known as the sense strand, has the correct genetic information in it, since the other one is different due to the base pairing rules. The message is single-stranded. The second thing is that the message is not DNA but a closely related molecule called RNA. This need not worry you because RNA still has four bases and the principle of carrying information in their sequence is the same. The copy is, for obvious reasons, known as messenger RNA or mRNA. The gene acts somewhat like a printing machine turning out multiple copies of mRNA which exit the nucleus and attach to ribosomes. A ribosome reads the mRNA and puts together amino acids according to the instructions in the message. When it comes to the end of the mRNA, the protein chain and the mRNA fall off the ribosome, which is then free to accept any other mRNA.

How are amino acids represented by the base sequence of mRNA?

This brings us to the very centre of the living process. We have the mRNA carrying DNA's instructions on the amino acid sequence of a protein to the ribosome. RNA has a totally different chemical structure from that of amino acids. There is no way in which an amino acid can directly relate to the messenger. The problem is comparable to having a message in a totally different language in hieroglyphics and having to translate it into English. The ribosome has to 'know' how the mRNA represents amino acids. We know that there are four different bases and there are twenty amino acids to be represented. A one-letter code could represent only four amino acids so that is out. A two-letter code could allow 16 different combinations (4×4); still not enough. A three-letter code could represent $4 \times 4 \times 4 = 64$ different triplets. Francis Crick, working on genetic mutants of bacteria, established that life uses triplets of bases to represent amino acids. The genetic code is a table of triplets that specifies which amino acid is specified by which triplet.

I need to explain one little quirk of RNA at this point. RNA has four bases, but instead of T it has a very similar one called U (for uracil). (There's a simple reason for DNA having T instead of U. If you are interested see the optional reading in Box 6.1.)

Box 6.1: Why does DNA have T while RNA has U?

It is all to do with DNA repair. DNA constantly suffers from damage, which can cause mutations with serious consequences. Cells do all they can to minimise these by repairing the damage. One cause of damage is that the base cytosine (C) is chemically somewhat unstable and tends to convert spontaneously to uracil (U). Cells have a repair system to deal with that; it removes the improper U and replaces it with C, thus fixing the problem. However the DNA repair system would not be feasible if U occurred normally, for it could not then distinguish between a 'proper' U and a one resulting from conversion of C. So U does not occur normally in DNA. Thymine (T) is simply U with a small chemical marker (a $-CH_3$ group). The repair system does not replace T. The use of T, to put it more simply, is that it says to the repair system 'I am really a U but it is normal for me to be here'. It is much less important in the case of RNA, for this is not repaired. Messenger RNA has a short lifetime so it does not matter if the occasional mistake exists. It is far more important for DNA in the genes to be as correct as possible.

Uracil (U) has identical base-pairing properties to thymine (T). Messenger RNA therefore has U wherever you would expect a T. A young American, Marshall Nirenberg, made an astonishing discovery which enormously advanced the elucidation of the genetic code. He found that a simple polymer of U, available off the shelf, acted as a mRNA. When translated by ribosomes a polymer of a particular amino acid called phenylalanine is produced. The only possible triplet in poly U is UUU. A triplet of bases specifying an amino acid is called a codon, so the codon for phenylalanine is UUU. This established the first break in deciphering the genetic code. It started a race between workers to decipher the entire code, which was quickly achieved. There was a surprise. There are only twenty amino acids used in protein synthesis so it was natural to suppose that twenty of the 64 possible triplets would be used and the others ignored. In fact, every triplet is used. Three are used as 'stop' codons; these signal to the ribosome that it is the end of the message, and protein synthesis is stopped and the protein chain released. All the other codons specify amino acids. This means that some amino acids have several codons. This may well puzzle you, for it seems at first sight to be an unnecessary complication. However it is important that all triplets mean amino acids, excluding the three stop triplets. A 'nonsense' triplet could jam the translation of a messenger. The code is universal to all forms of life. To be clear on this, if you make a list of codons and opposite it a list of the names of the amino acids each codon represents, the lists are identical for all life forms; they all interpret codons in terms of amino acids in the same way from bacteria to humans (slightly qualified in a rare few instances which do not change the overall concept).

As mentioned, there are no structural or chemical reasons relating a given amino acid to its codon. The genetic code is what Crick called a frozen accident. It presumably just happened that way in the very early stages of life from which all subsequent life forms evolved. Once the code was established it could not be changed, for changing even a single codon would cause havoc in protein synthesis. The ribosome has to start at the first codon at the beginning of the mRNA and move along one codon at a time. Incidentally this means that if a mutation during DNA replication misses out a single base, the reading of the rest of the message would be erroneous and the protein produced would be garbage. For example, if the first two codons were UUU CCC, missing out one U would give a reading of UUCCC – and everything after this would be different from what would give rise to the correct protein.

The synthesis of a protein is summarised as below.

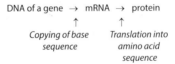

DNA of a gene → mRNA → protein
             ↑   ↑
      Copying of base   Translation into
         sequence   amino acid
                           sequence

There remains a major problem to be explained. How do so relatively few genes code for so very many proteins? The answer is that the string of bases comprising a gene includes the equivalent of punctuation or break points. The RNA that is copied from the gene includes all the punctuation, but it is then chopped at these break points and selected pieces are spliced back together to code for the particular protein that is needed. The mRNA that leaves the nucleus is therefore an edited copy of the information in the gene. The instructions on where to cut and splice the mRNA appear to be coded in what was once thought to be 'junk' DNA – DNA that does not code for proteins. (See Chapter 7.)

How are the codons translated by the ribosome into a protein chain of defined amino acid sequence? Somehow the amino acids and their codons must in some way recognise each other. I think it was Crick who first reasoned that since the codons are triplets of bases, then hydrogen bonding would be involved.

The breakthrough came when a new type of RNA molecule was discovered – called transfer RNA or tRNA. There is at least one tRNA molecule for each of the twenty amino acids. They act as adaptors to associate amino acids to their codons.

The function of tRNA is very similar to that of adaptors for mobile phone chargers and suchlike when you travel to a country whose electric sockets do not accept the home plug. An adaptor accepts your razor plug at one end and the other end fits the foreign socket. In protein synthesis there is at least one type of adaptor (tRNA) for each of the twenty amino acids involved in protein synthesis. The essence of the mechanism is that the adaptor fits to its codon by A–T, G–C base pairing, and at the other end accepts its amino acid. This lines up amino acids in their correct sequence ready to be linked up to form a protein (Box 6.2).

Gene mutations affect the amino acid sequences of proteins, causing genetic diseases

A gene mutation is, in its simplest form, an incorrect base in the sequence of the DNA. This leads to an incorrect

Box 6.2: A deeper look at the translation of codons into amino acid sequences

There is at least one adaptor RNA molecule for each of the twenty amino acids. Multiple copies of each of these are present in cells. Let us take the amino acid glycine as an example. Its codon is CCC on the mRNA. The tRNA for this amino acid has at one end a site to which a specific enzyme attaches glycine and glycine only – not any other amino acid. There must be no mistake in this or the wrong amino acid will be inserted into the protein. At the other end there is a triplet of bases called an anticodon. In this case it is GGG. Remembering that G hydrogen-bonds with C, you can see that GGG will bind to CCC. Thus, Watson–Crick base pairing automatically attaches glycine to its codon for this amino acid. The same applies to all twenty amino acids, each having its own tRNA and an enzyme that attaches the appropriate amino acid to it. The ribosome moves along the messenger and links the amino acids together until it reaches a termination codon, where the protein chain is ejected. Its amino acid sequence is that specified by its gene.

You can see from this that Watson–Crick base pairing really is the foundation of life. I will remind you that the pairing also provides a method of accurately replicating DNA – all 6.4 billion bases per cell in

Box 6.2: (Continued)

humans. It is the basis of reproduction. It is almost incredible that all life on this planet is based on the simple fact that A pairs with T and G pairs with C.

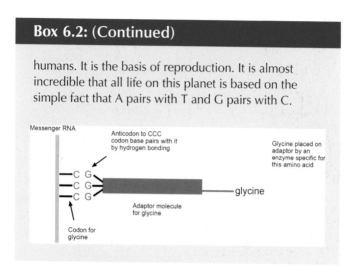

codon in the mRNA which the ribosome translates into an incorrect amino acid in the protein synthesised. To return to the analogy of a protein being a very long word made of twenty letters, an incorrect amino acid resembles a single spelling mistake in the word. The effect of this in a protein can vary from death of the embryo to a genetic disease in the offspring, or, if it is in a non-critical place in the chain, it has no effect. The hereditary disease cystic fibrosis is caused by one particular protein lacking a single amino acid out of a chain of almost 1500. Pre-implantation genetic diagnosis (Box 6.3) is a clinical service offered to couples with a known familial genetic disorder, such as cystic fibrosis, and who wish to prevent that disorder being passed on to a child.

What causes mutations? The most obvious cause is errors made when DNA is replicated. There are 6.4 billion bases to be copied and some mistakes are inevitable. The mechanism is remarkable in the strategies used to minimise them. For example the enzyme which links the nucleotides together has a proof-reading ability. After adding each new base it checks that it is correct. If not, it removes it and gives a chance for the correct one to come in. It sounds impossible for a molecule of protein to do that but an automatic mechanism achieves it.

There is also an elaborate system of checking the finished DNA for mistakes; when found, the erroneous

Box 6.3: Pre-implantation genetic diagnosis

The service of pre-implantation genetic diagnosis relies on technology to detect specific genetic characteristics in embryos created using IVF (in vitro fertilisation) before they are implanted into the uterus. Conditions that may be detected by preimplantation genetic diagnosis include: sex-linked disorders, such as haemophilia, fragile X syndrome and neuromuscular dystrophies; single gene defects, such as cystic fibrosis, sickle cell anaemia and Huntington disease; and chromosomal disorders, such as Down syndrome. To perform the procedure, eggs and sperm collected from the prospective parents are combined in the laboratory to allow fertilisation to take place. If successful, cell divisions in the embryo commence. After several days, when approximately 6–10 cells are present, a single cell is removed. The removal of one cell does not seem to damage the health of the embryo at this early stage. The single cell is processed as rapidly as possible to perform all required genetic tests. Because all the cells in the embryo are identical, testing one will reveal the genetic identity of all. If the tests show that the cell

Box 6.3: (Continued)

does not contain the genetic disorder(s) under consideration, the entire embryo is considered to be healthy. The embryo is then implanted into the uterus in the hope that it will survive to initiate a pregnancy.

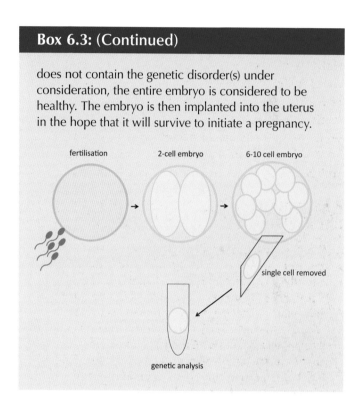

part is removed and the gap repaired correctly. Spontaneous chemical changes in the bases occur all the time, and these are also repaired. Such mechanisms keep the number of errors to a level compatible with life, but some mistakes do happen and genetic diseases result from uncorrected mutations. When a repair system is deficient, problems arise. For example, in the genetic skin disease xeroderma pigmentosum, DNA damage caused by exposure to UV light in sunshine causes cancerous skin lesions.

Box 6.4: Still a way off beating cancer

Despite the USA's President Nixon declaring a 'War on Cancer' in 1971, statistics show that only very minimal progress is being made against this spectrum of disease. While some cancers are slowly declining, the rates of death due to other forms of the disease – such as cancers of the liver, pancreas and uterus – even increased during recent periods. Rates of cancer attributed to the human papilloma virus (HPV) are also creeping up despite the existence of the HPV vaccine (Chapter 4, page 82).

The numbers speak of the incredible degree of complexity that exists in trying to manage what goes wrong when cancers first develop, and in preventing their metastatic spread to other parts of the body. Developing and rolling out effective and affordable treatments is also an ongoing battle.

Pioneering scientist Dr James Watson – one of the duo awarded a Nobel Prize for determining the double helix structure of DNA in 1953 – published an opinion paper in 2013 that outlined his thoughts on where science and medicine were going wrong in attempting to beat cancer. He claims 'The main factor holding us back from overcoming most of metastatic cancer over the next decade may soon no longer be lack of knowledge but our world's increasing failure to intelligently direct its "monetary might" towards more human-society-benefiting directions.'

Watson also considers that some of our approaches relating to preventing cancer – such as the belief (see page 82) that increasing dietary antioxidants like

Box 6.4: (Continued)

β-carotene, vitamin A, vitamin C, vitamin E and selenium can prevent the disease – are flawed. He believes a concerted and organised effort is required to better direct funding for cancer research towards common and simplified goals across the world.

Another important source of genetic damage is the destructive effect of a reactive form of oxygen called oxygen free radicals. The use of oxygen in the liberation of energy from food such as sugar increases the yield of ATP 16-fold to 32 molecules of ATP molecules produced per molecule of glucose (see Chapter 2). But there is a darker side. Oxygen itself is a relatively benign molecule and when converted to water during the oxidation process it is completely safe. But in the body, occasionally a single electron is added to oxygen instead of the two needed to form water. This produces an extremely reactive form known as superoxide which attacks a covalent bond of some other molecule to obtain an electron as a partner for the single electron. The attacked molecule itself now has a single electron and in turn it attacks another molecule. Long chains of random destruction of biological molecules, including DNA, occur. Reactive oxygen species are believed to be involved in ageing and many other deleterious effects.

The body has protective mechanisms against these oxygen free radicals. One is the enzymic destruction of reactive oxygen species by converting them to water; the second method is by antioxidant compounds which limit the damage caused. They are themselves attacked, but when that happens they terminate destructive chains of reaction set up by the renegade reactive oxygen atoms because, after being attacked, they do not then attack other molecules. Several biological molecules are antioxidants, such as vitamin C, vitamin E and β-carotene. The value of eating vegetables is partly that they contain antioxidants (but see Box 6.4).

7

The life and death of cells

Most forms of life consist of single cells not connected to others. Bacteria are an example. These are self-contained organisms that perform all the processes needed for their growth and reproduction. Their survival strategy is to reproduce as rapidly as possible and so outgrow rival organisms. An *Escherichia coli* bacterium cell can, under optimal conditions, reproduce itself in approximately 20 minutes. If a single cell and its offspring could continue to multiply at this rate it would form a mass of cells equal in volume to that of the Earth in about three days. (Even for bacteria life is not easy, so this does not happen.)

Bacteria do not have a cell nucleus; their DNA is simply in contact with the rest of the cell. (Cells like this are called prokaryotes in case you meet the term elsewhere.) In so-called higher cells (called eukaryotes), the DNA is confined to the cell nucleus, a microscopic sphere surrounded by a membrane (see Fig. 2.2). Yeasts are single cells of this type and are more complex than prokaryotes, but evolution has produced much more complex organisms in which large numbers of cells collaborate together. Humans have about 10 trillion cells.

All cells have in common that their genetic material is in the form of DNA, and all are surrounded by a membrane that is essentially the same in all cells. Although there are superficial differences between DNA function and replication in bacteria and humans, basically they are the same. Their DNA, and genes, are much the same in function. The same is true for the way in which release of energy from food and its utilisation is achieved. They all make ATP, which is the immediate source of energy for virtually everything in life (see Chapter 2). (It is worth noting that viruses are not cells and some of them have RNA instead of DNA.)

Strict control of cell multiplication is necessary in animals

In higher cells of the type found in animals and plants, replication is slower than in bacteria, taking typically about 20 hours, and it is strictly controlled. A cell in your body is allowed to divide only if it has 'permission' to do so. This is given in the form of special proteins known as growth factors produced by surrounding cells, though how this system of 'governance' works is not fully understood. The cells of a human must do only what is in accordance with the needs of the body as a whole.

Occasionally a renegade cell starts multiplying out of control, without permission as it were, and may then mutate and escape to other parts of the body – a process known as metastasis. This is what a cancer is – a breakdown in the growth control of a cell (see Box 7.1). Benign tumour cells also multiply uncontrollably but are much less dangerous because they do not break loose and spread to other parts of the body.

Box 7.1: Leukaemia results from uncontrolled cell division

Haematopoiesis is the process by which blood stem cells residing in the bone marrow and a few other sites divide and mature along defined pathways to produce all of the different types of cells in our blood. Under normal conditions, some of the self-renewing stem cells form into different types of cells called blast cells, which then divide and differentiate further to generate mature cell types: white blood cells (for immunity), red blood cells (which carry oxygen), and platelets (for clotting). Each step in the process is highly regulated by key genes being switched on and off, and specific growth factors stimulating growth and differentiation. Occasionally the controls that guide haematopoiesis fail. An individual population of blast (A) or mature (B) cells escapes normal regulation and multiplies abnormally. This condition is known as leukaemia.

As a result of the leukaemia, the normal balance of cells in the blood becomes disrupted. With one population of cells being preferentially produced, other cells become relatively low in number. Reduced levels of red blood cells and platelets cause symptoms such as

Box 7.1: (Continued)

anaemia, tiredness and slow blood clotting.
Vulnerability to infection also increases, as the majority
of the circulating white blood cells are too immature or
have abnormalities that prevent them from performing
normal immune functions. If treatment fails to rectify
these imbalances, the patient may die. A diagnosis of
what type of cell is involved in the leukaemia – and
hence which treatment is appropriate – can be made by
studying samples of blood and bone marrow cells taken
from the patient and viewed under the microscope, and
analysing them for chromosome and cellular
abnormalities.

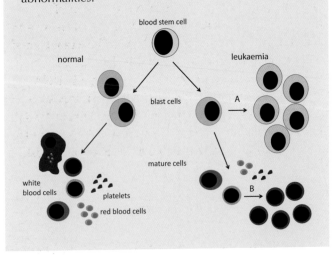

DNA replication in animal and plant cells has the
curious feature that it cannot copy the terminal sections
of chromosomes. If nothing were done about this it
would mean that every time a cell divides the new DNA

molecules of the daughter cells would be shortened – a potentially disastrous situation since genes would gradually be deleted. The solution adopted is that the ends of the 'real' DNA are lengthened by a stretch of 'phony' DNA, a few hundred nucleotides long, added separately one nucleotide unit at a time. This, as it were, is sacrificial DNA – it has no genetic information in it. It is shortened at every cell division but as nothing is lost of the 'real' DNA there is no loss of genes. These protective ends, known as telomeres, play an important part in that they are also cell division 'counters'. When the telomeres become shortened to a critical length, the cell cannot divide any more. It seems that each cell is given its ration of telomeres which are tickets for a set number of divisions, and that is it. There is evidence that, as we age, telomeres get shorter. Cancer cells reproduce rapidly and without apparent limit. To allow this they have an enzyme that continually replaces lost telomeric DNA. Bacteria don't have this problem because their DNA is circular, without ends to shorten. Recently an Australian, Elizabeth Blackburn, and two Americans, Carol Greider and Jack Szostak, received a Nobel Prize for their work on telomeres (Box 7.2).

Stem cells

The cells of different organs of the body are specialised, meaning they have different forms to suit their function. There are liver cells, brain cells, muscle cells and so on. Liver cells perform biochemical tasks that do not occur in skin cells, and vice versa. These specialised cells are

Box 7.2: Telomeres – from Tasmania to the Nobel Prize-winners' stage

Telomeres act as protective caps at the ends of chromosomes in cells, ensuring that valuable DNA is not lost or damaged during replication. By way of visual analogy, they are often described as the cellular equivalent of the plastic tips that prevent shoelaces from fraying. Telomerase is the enzyme that repairs telomeres to ensure their length remains adequate to serve its purpose.

The place of telomeres and telomerase in understanding fundamental cell biology and human health is considered to be so important that it attracted a Nobel Prize. One of the awarded scientists was Elizabeth Blackburn.

Blackburn was born in Tasmania in 1948, the second of seven children. After a childhood filled with watching nature and animals, she gained an Honours degree in biochemistry at the University of Melbourne. A subsequent year of Masters research on amino acid metabolism with Frank Hird gave her the experience to qualify for PhD studies at Cambridge University with Fred Sanger. Blackburn's doctoral work focused on the application of Sanger's and other colleagues' new methods for sequencing RNA and DNA. At the time, this was absolutely cutting edge.

In the mid 1970s Blackburn moved to Yale University in the United States, where she conducted her postdoctoral studies. Here she began studying telomeres in earnest, working to sequence the DNA at the ends of chromosomes in a single-celled organism. Working with Jack Szostak, Blackburn deduced that

Box 7.2: (Continued)

there must be an enzyme which operates to restore the telomeres each time a cell divides.

After moving to University of California Berkeley, and working with graduate student Carol Greider, Blackburn discovered this enzyme – telomerase. She now continues to work on telomeres and telomerase, and in particular is focused on determining whether shortened telomeres or inadequate telomerase function could contribute to ageing and ill health.

The Nobel Prize in Physiology or Medicine 2009 was awarded jointly to Elizabeth Blackburn, Carol Greider and Jack Szostak 'for the discovery of how chromosomes are protected by telomeres and the enzyme telomerase'.

known as somatic cells. When they divide, they can produce only their own type. Liver cells produce only liver cells and skin cells only skin cells, and so on.

All of the different cell types arise from a single fertilised egg cell. As the embryo develops, it multiplies into a mass of unspecialised cells known as embryonic stem cells. These are capable of giving rise to all the different cells of the body and for this reason are described as pluripotent. When stem cells divide, of the two daughter cells one can form a somatic cell while the other can stay as a stem cell so that a pool of stem cells is retained. The change from pluripotent stem cells to somatic cells is called differentiation. This is normally irreversible in the body. As the biological clock moves on from stem cell to differentiated somatic cell, only those genes appropriate to the function of the tissue they belong to are switched on. Liver cells for example do not produce the contractile proteins of muscle.

There is great interest in stem cells because their ability to form into different types of cells hints at their potential to repair damaged organs. Embryonic stem cells in principle are technically the easiest to use in this role: since they can more easily be multiplied in the laboratory without losing their pluripotency they therefore could be used to produce any type of somatic cell. However they cannot be used in humans because of ethical objections; their initial preparation involves destruction of human embryos surplus to requirements in the IVF technology. Another objection is that there are risks involved: such cells derived from individual A would be genetically different from individual B (the potential recipient) and would face immune rejection, since the genes present in the stem cells and their descendants would give rise to proteins foreign to the patient. The immune system may produce antibodies against these and cause organ rejection.

An alternative solution to these difficulties might be to use adult stem cells. Most tissues of the body retain a small pool of stem cells, whose function is presumably to maintain tissues with new cells as needed. Adult stem cells are different from embryonic stem cells in that they can only give rise to a limited range of cell types normally found in the organ – their differentiation might be regarded as incomplete. For example the adult stem cell precursors of blood cells can produce only various types of blood cells in their final differentiation step.

If adult stem cells could be used on the patient from whom they were collected there would not be ethical objections or the problem of immune rejection. There is the difficulty, however, that relatively small numbers are present in adult tissues and they do not multiply so easily in the laboratory. Nonetheless the method has been successfully used in certain cases and much work is being done to exploit their therapeutic potential.

One new development is attracting great interest. Once a pluripotent stem cell has converted to a specialised somatic cell the process is never reversed in the body (in vivo) – a liver cell cannot go back to being a pluripotent stem cell. But it has been found that some ordinary somatic cells, such as a type of skin cell, can have their biological clock turned back in the laboratory (in vitro) to pluripotent cells resembling embryonic stem cells. This might seem to offer the ideal situation because a patient's own cells might then be used to repair damaged tissues. The conversion to stem cells is done by introducing certain embryonic genes involved in mRNA synthesis (transcription factors) into the somatic cell. This is done using a virus to transport the genes into the cells, which raises concern that the procedure might alter the genetic make up of the cells and the worry that the cells may turn cancerous. Much research on this is being done, because if these induced pluripotent stem cells could be used safely on patients it would be a major medical advance.

Programmed cell death

Controlled cell death is more important to multicellular animals than you probably realise; the mechanisms involved are almost as elaborate as some of those necessary for life. Most somatic cells in your body have a built in self-destruct mechanism waiting to be triggered. The process is known as apoptosis (Box 7.3), a word derived from Greek; *apo* means 'detached' and *ptosis* means 'falling' in the sense of falling leaves. It is usually pronounced as a–poptosis, though by the original discoverer as apo-tosis. Apoptosis is different from simple cell death (necrosis, which can be a messy business); it involves destruction of DNA but the cell is neatly taken apart in pieces which are engulfed by white cells, and the components are recycled. The point of it all is to protect the body from some hazards which could destroy it. The self-destruct system can be triggered by several means. One of them occurs when division of a cell is stopped because it has damaged DNA. Allowing it to proceed to cell division could result in the damaged DNA causing the cell to be cancerous. A protein named p53 detects the damaged DNA and orders the cell to self-destruct. The mechanism of this is surprising in that the p53 protein causes the leakage from mitochondria of a protein (cytochrome C). In the mitochondrion, cytochrome C is a highly respectable well-behaved enzyme which has been known for decades as essential in ATP generation during food oxidation. However, when it is released from the mitochondria, it is a death sentence for that cell.

The gene for p53 is known as a tumour repressor because it protects against cancer by ordering apoptosis

Box 7.3: Apoptosis is controlled cell demolition

Apoptosis is a highly controlled process through which cells initiate their own death and clearance from the body. In an article written for the journal *Nature Reviews*, authors Rebecca Taylor, Sean Cullen and Seamus Martin put it this way:

'In many ways, what happens during apoptosis is akin to how large buildings are demolished to make way for new developments. During demolition, it is important that the process is carried out in a safe and controlled manner to ensure that neighbouring structures remain unaffected. To achieve this, a specialised demolition squad is called in and, all being well, these experts carry out the task in a precise and highly efficient manner. After demolition has been completed, the debris is removed and a new structure takes the place of the old one within a short time.' (Reprinted by permission from Macmillan Publishers Ltd: *Nature Reviews*. Taylor RC, Cullen SP, Martin SJ (2008) Apoptosis: controlled demolition at the cellular level. *Nature Reviews* 9(3), 231–241.)

As well as the apoptotic cell, a special type of white blood cell, known as a phagocytic cell, is an active participant in the process. As shown in the diagram, four simplified steps take place in apoptosis. (1) Identification: the apoptotic cell releases soluble molecules that trigger the movement of the phagocytic white blood cell towards it. (2) Targeting: the apoptotic cell expresses molecules on its surface that match up with receptors on the phagocytic cell. (3) Engulfment: changes take place in the phagocytic cell that allow it

Box 7.3: (Continued)

to form a cup around the target apoptotic cell. (4) Destruction: the apoptotic cellular corpse is fully swallowed by the phagocytic cell and its components are broken down for recycling.

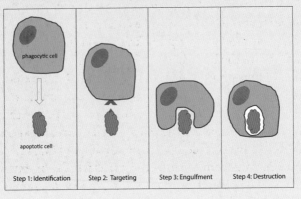

| Step 1: Identification | Step 2: Targeting | Step 3: Engulfment | Step 4: Destruction |

of abnormal cells. Such a system is obviously fraught with the danger of the self-destruct mechanism being activated erroneously, so there are controls against this. Nature tends to favour the push-pull system of opposing controls to achieve control stability. There are control proteins, one type promoting apoptosis and the other opposing it. These normally are correctly balanced but imbalances can occur due to gene mutations. If the anti-apoptosis proteins are in excess, then cells that should be destroyed may be allowed to survive, with the risk of cancer developing.

Another role of apoptosis is to protect against virus infections. Viruses in the bloodstream are destroyed by phagocytic white blood cells which engulf and destroy them. But when viruses infect host cells of the body they go right inside the infected cells, where they are immune to this form of attack. The phagocytes can't get at them inside the cells, and there the viruses reproduce, with the danger that the infection will escalate. The thymus is a gland that produces what are known as killer T-cells. These roam the body looking for virus-infected cells to kill and thus prevent virus multiplication. But how does the killer T-cell know which cell is infected? Inside all cells there are enzymes which chop up 'samples' of the proteins inside the cell into small pieces and display them on the *outside* surface. The pieces on display include those derived from the infecting virus which has entered the cell. The killer T-cells are alerted to look for somatic cells that display viral peptides and attack them. Somatic cells have special death receptor patches on their surface through which the killer T-cells deliver the death sentence.

Apoptosis occurs normally on a large scale. As one example, in the immune system untold millions of cells are signalled to self-destruct in the bone marrow where they are formed. These are immune cells which would attack the animal's own cells if released from the bone marrow. Apoptosis also is important in the development of embryos. When a hand or foot is forming, it is a bit like a flipper with webbing or membrane between the fingers or toes. The separation of fingers and toes is due to the cells in the webbing between them self-destructing.

The human genome project gave startling information about human DNA

In the last couple of decades a new era has opened in discoveries of how life works, due to new methods in manipulating DNA, and the amazing advances in computer processing which enabled the rapid sequencing of DNA.

Fred Sanger of Cambridge was awarded his second Nobel Prize for developing a method for sequencing DNA; this means determining the sequence of the A, T, G and C bases on the DNA which, as described, carries the information for the amino acid sequence of proteins. His first Nobel Prize was for sequencing the protein insulin. Sanger spent his life working personally in the laboratory. His research group was small. He surprised many at the age of 65 by retiring from his research. On his last day, he finished his laboratory work, had a cup of tea with his colleagues, and rode his bicycle to his home on the outskirts of Cambridge, where he and his wife lived. After having twice had such a tremendous impact on the most fundamental aspects of life – the structure of proteins and a method for sequencing DNA, both of which were landmark achievements in biological science – he turned his interests to enjoying a boat on an English river and growing roses.

Sanger's method of DNA sequencing raised what seemed to be a distant dream, namely to sequence the bases in an entire human genome. Genome is a collective term meaning all of the DNA of an organism. But because there are 3.2 billion base pairs in human DNA, this was regarded almost as an unattainable goal. However, many years after Sanger's retirement, using automated procedures and cooperation between different laboratories, the task was completed in around 3 years at a very great cost. The entire blueprint of the sequence of all the 6.4 billion bases of double-stranded human DNA is now available. When the structure of human DNA became available, it caused shock because it looks like a disorganised mess thrown together haphazardly. One distinguished worker in the field likened it to a bedroom, refrigerator or toolshed which had not been tidied up for a long time. The genes seem to be scattered around any old how and, surprisingly, the protein-encoding genes occupied only a small fraction of the total DNA – about two per cent. The rest was called 'junk DNA' – presumed to have no function. However, we now think that this 'junk DNA' is actually quite important, because it houses numbers of mini-genes producing small RNA molecules. These are not mRNA coding for proteins, but they appear to control the protein synthesis directed by genes (Chapter 6). It may be that the mini-genes have been important in the evolution of complexity in organisms. One would have thought that increasing complexity would be due to increased number of protein-encoding genes but the facts don't correlate well with this. For example, a very simple small nematode roundworm has 18 000 genes. Humans certainly have more – approximately 20 000 to 25 000 – but the nematode has a total of only 1000 cells in its entire body as against about 10 trillion cells in a human. The mini-genes can prevent expression of conventional protein-coding genes by preventing mRNA translation; this has considerable medical potential.

When the base sequence of the bacterium *E. coli* was determined, it was found that there is little or no junk DNA; the genes are arranged neatly with relatively little space between them. Altogether the bacterial genome looks much tidier than the human one.

Revolutionary sequencing methods have advanced to such an extent that it now takes only a short time to sequence an entire human genome. It seems that the entire DNA sequence of individuals could be available to patients at affordable cost ($5000 in 2013). This could have benefits in, to give a single example, treatment of complex diseases such as cancer in individual patients

Box 7.4: More than just the genes – epigenetics

It's a well-accepted view that both genes and the environment influence the appearance, capabilities and health of each person. Genes code for certain protein sequences, and determine features such as whether our eyes are blue or brown, if we're innately good at short- or long-distance running, and whether we're likely to develop baldness or diabetes as we age. Environmental factors like sunshine, diet and exercise exert an additional effect, shaping characteristics such as darkening of skin, body fat, and trained muscle bulk.

Box 7.4: (Continued)

But did you know that environmental factors can also have intergenerational impacts? A study conducted in Sweden showed that if a father did not have enough food available to him during a critical period in his pre-pubertal development, his sons were less likely to die from cardiovascular disease. Something that moderated the action of genes was being passed on to the sons of those particular fathers.

This 'something' is known as epigenetics. Epigenetic changes can switch genes on or off, and determine which proteins are transcribed.

There are three systems that are thought to be involved in epigenetic control of gene expression:

1. *DNA methylation*
   This is a chemical process that attaches a methyl group to DNA. The methyl group prevents that section of DNA being transcribed for protein production.
2. *Histone modification*
   Histones are special proteins that act as a spool around which DNA can wind – a spooled section of DNA cannot be transcribed for protein production.
3. *RNA-associated silencing*
   Some sequences of RNA bind to DNA and prevent transcription, or trigger DNA methylation and histone modification.

Epigenetic changes are critical for normal development, but can also go wrong and trigger disease states such as cancer and intellectual disabilities.

whose treatment might be more precisely targeted to the individual case. There is still a great deal yet to be understood on how the genome works as a whole, though how individual genes work and are controlled is largely understood. One of the least understood aspects of DNA function is that certain gene controls are exerted without any change to the base sequence of the gene DNA. This is known as epigenetic control – sort of a gene control system on top of 'ordinary' gene control (Box 7.4).

8

The origin of life

The Earth was formed approximately 4.5 billion years ago. It took about a billion years to cool down sufficiently for the surface temperatures to be compatible with the existence of life, so life is unlikely to have started until after about 3.5 billion years ago. There is evidence from fossil-like structures in dated rock sections that indicates that life forms resembling bacteria existed approximately 3 billion years ago. Although we tend to regard bacteria as primitive life forms, in fact they are in a molecular sense highly complex. If you consider the bacterium *E. coli*, it is a membrane-coated cell, with a DNA-based genetic system. The way DNA is replicated and proteins synthesised in *E. coli* is, in a molecular sense, virtually the same as happens in your own cells. So bacteria cannot be regarded as being candidates for the most primitive forms of life. Viruses at first sight might seem to be likely candidates for this, but they probably should be regarded in a molecular sense more as advanced parasites. They are simple in that many consist of little more than a molecule of nucleic acid inside a protective coat, but they can exist only by hijacking a cell and using its molecular apparatus for their own ends. For that matter it is debatable whether viruses should even be regarded as living organisms. Some can be crystallised like chemical molecules, and on their own, outside of host cells, they show no signs of life.

Life must have arisen between the time Earth had cooled enough and the date of the earliest fossils of bacteria-like organisms. It seems that prokaryotes (bacteria) dominated life for 2 billion years. Eukaryotes (higher-type animal and plant cells) did not appear until about a billion years ago. All life forms are basically so similar that it is believed that there was a single origin of life, or if there were more than one original form, only one has survived.

A question here is what we mean by life. Initially it would have to be a collection of molecules that could reproduce to form replicas of itself. At this stage it must have been a simple collection of relatively small molecules that would not leave any fossils, so we may never know what the earliest form of life was.

We can only speculate on the origin of life. There was no one around at the time to witness it; there is no direct evidence about the first form of life. Nevertheless there is a lot of information relevant to the early stages of life.

Where did the first biological molecules come from?

The first question relates to the question of where life got its components from. There were no enzymes to form them and at first sight it might seem unlikely that components of cells could have formed spontaneously. But in 1952 a young PhD student, Stanley Miller, working with Professor Harold Urey at Chicago University, threw a flood of light on this. Miller decided to see if biological compounds could have been formed on the primitive Earth. He assumed the atmosphere of Earth around 4 billion years ago probably contained hydrogen, water, ammonia and methane. He built an apparatus in which these components circulated continuously through a closed tube. The circulation was driven by the inclusion in the system of a boiling flask of water, and the gases were passed through a larger flask through which electric sparks were fired, to simulate lightning. After passing through this, the water vapour was condensed and the water returned to the boiling flask to be recirculated. At the end of a couple of weeks he collected the water and analysed it for biological molecules. Apparently he had been warned against doing this project because it seemed likely that a meaningless mixture of a huge number of different molecules would be produced. Instead, three amino acids present in all present-day proteins were present. The important point is they were produced in relatively large amounts, not as just trivial amounts out of hundreds of others.

Since the Miller–Urey experiment, others have followed up the work using varied gas mixtures (since the actual composition of the primitive Earth's atmosphere is not known; but this proved to be not critical to the experiment). A very striking experiment showed that if hydrogen cyanide was present, large quantities of adenine (the A of ATP and one of the bases in DNA) were formed. If formaldehyde was included, the sugar ribose, present in ATP and RNA, was formed. Formaldehyde and cyanide are simple molecules that could be expected to have occurred on primitive Earth. The results showed that a large number of compounds present in living organisms could have been produced on primitive Earth. It seems that the supply of such molecules would not have been a major problem in the establishment of life.

The ancient RNA world

It seems that the early forms of life used RNA, not DNA, as their genetic material. All cellular life has DNA as the carrier of genetic information. But apparently it wasn't always like that: all life once had RNA to do this and DNA did not exist. The reason for this is probably that RNA is chemically easier to make than is DNA. In modern life a complex enzyme system is need to convert the sugar ribose to deoxyribose, even though it requires only the removal of one oxygen atom, while ribose itself is made in Miller–Urey type experiments.

The probable existence of an ancient RNA world is generally accepted and there are still traces of it to be found. It could well be that the use of mRNA in the crucial business of protein synthesis is a leftover from the

RNA world. Two workers in America (Nobel Prize recipients Thomas Cech and Stanley Altman) separately shook the biochemical world with a discovery that was difficult to believe. Until then, it was accepted that all biochemical reactions are catalysed by proteins, but these workers found that, in a few rare instances, reactions were catalysed by RNA molecules. These are not as efficient as protein enzymes, but the very fact that they worked at all was startling. It seemed that the examples found were fossil-like leftovers from the ancient RNA world.

The discovery gave a likely solution to a problem that dogged thinking on the origin of life. Proteins are made by ribosomes directed by genes, and ribosomes are a mixture of protein and RNA molecules. There is a chicken-and-egg problem. Which came first, RNA to direct protein formation, or proteins to carry out reactions? Then another startling revelation came. One of the most important reactions in life is that which links amino acids together to form protein chains. This occurs in ribosomes and it was always assumed that it was done by an enzyme. The startling discovery was that it is actually catalysed by a molecule of RNA in the ribosome, not by a protein molecule. This looks like a major fossil hangover from the ancient RNA world. A concept now is that in those early days ribosomes were entirely RNA and that proteins were added later to increase efficiency. From this it would be possible for RNA to act both as the genetic material and to catalyse reactions. It removes the chicken-and-egg problem (Box 8.1).

Box 8.1: Early life looked a lot like RNA

There is a growing body of scientific evidence supporting the idea that early life could have been based on RNA.

In life forms as we currently know them, charged atoms, or ions, of the element magnesium are essential for RNA folding and catalysis. However, magnesium was scarce billions of years ago when life evolved on Earth, and oxygen gas was essentially non-existent. A group of American scientists found that under iron-rich, oxygen-free conditions, RNA can act as an enzyme and catalyse single electron transfer.

Ancient RNA probably also relied on vitamins to catalyse chemical reactions. Canadian scientists identified an RNA molecule that can use vitamin B1 to decarboxylate (i.e. remove a carbon dioxide molecule) another molecule.

Together, the two studies provide evidence that under conditions found on Earth billions of years ago, RNA could have facilitated transfer of energy and metabolism of simple materials in a manner consistent with establishing self-replicating life.

Why has RNA been replaced by DNA in all cellular life?

RNA and DNA are very similar. As said earlier, the sugar in RNA nucleotides is ribose, while that in DNA lacks one oxygen atom and is called deoxyribose. The nucleotide chain is chemically more stable when the sugar is deoxyribose rather than ribose. This is important for, at all costs, the integrity of the DNA must be preserved, and probably this is the reason for DNA taking over the function of RNA from the ancient world. However there are several viruses, including influenza and the AIDS virus, which use RNA as the genetic material. The possible or probable reason why these can get away with having RNA rather than DNA is that the genomes of viruses are much smaller than those of cells and are therefore less likely to have spontaneous breaks than much longer ones. Also, the viruses are multiplied so rapidly that a few casualties are of no great threat to virus survival.

In Box 6.1 (page 112) the reason for DNA having the base thymine instead of uracil (as in RNA) is discussed.

How were cell membranes formed?

Whatever the first form of life was, it was presumably necessary at some stage for it to be contained by a membrane, for otherwise a simple collection of molecules would have been dispersed by a wave or rainstorm or whatever. All living cells are surrounded by a membrane made of a double layer of special fat molecules. All biological membranes basically have the same appearance in electron microscope images of cross-sections. The fat molecules in membranes have a special characteristic. They have two distinct ends; one is water-liking (hydrophilic), and the other end is a water-hating (hydrophobic) tail. Some of the fat in egg yolk is of this type. Lecithin, often mentioned in discussions of food composition, is a typical component of cell membranes.

It was difficult to imagine how the most primitive life forms could have produced what looks like a fairly elaborate structure, and yet it is difficult to see how life could have survived without a protective membrane to hold the primitive life forms together. The interesting discovery was made that if you simply shake compounds like lecithin in water, small membrane-bound structures called vesicles appear, roughly the size of cells, and in the electron microscope these are identical in cross-section to modern cell membranes. Their structure is shown in Fig. 8.1; the molecules automatically arrange themselves into a double layer with the hydrophobic tails in the middle away from water, and the hydrophilic part in contact with water on both sides of the membrane. This is a stable form of structure. It has given rise to the speculation that somehow the first self-reproducing

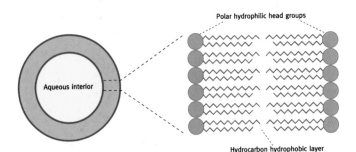

Figure 8.1: The structure of cell membranes. The components are a particular sort of fat molecule which at one end is water-loving (hydrophilic), represented by the spheres, and at the other end is water-hating (hydrophobic). Such molecules when simply shaken in water automatically assemble into a membrane of the form shown, in which hydrophilic parts are exposed to the water and hydrophobic parts are not exposed to it.

Box 8.2: Mitochondria offer a clue to our cellular history

The cells that make up humans, other animals and plants are known as eukaryotes: they contain a nucleus and other internal features called organelles. (See Fig. 2.2b.) Mitochondria, the energy-producing units of the cell, are one type of organelle. Bacteria and other cells that do not have organelles or a nucleus are referred to as prokaryotes.

Although it is difficult to know exactly how the characteristics of modern eukaryotic cells evolved, scientists have some ideas. For example, many believe that mitochondria were once free-living prokaryotes that were engulfed inside an ancestral cell type, and became permanent passengers. There are two main theories that describe how this may have arisen.

The first theory suggests that the original host cell was a eukaryote, but was not able to use oxygen to generate its own energy. The host cell acquired the mitochondria by an active process known as phagocytosis, in which the host cell wrapped itself around the mitochondrion and then completely enveloped it. The capacity of mitochondria to utilise oxygen to generate energy was advantageous to the host cell, and thus evolution favoured the relationship becoming permanent.

The second theory suggests that the original host cell was not a eukaryote, but instead a larger prokaryote. Although it is not clear how the mitochondria became internalised, the evolutionary driver behind the relationship may have been the value

Box 8.2: (Continued)

of mitochondria-produced hydrogen ions as a source of energy for the host. With further evolution, this prokaryote-inside-a-prokaryote then gradually formed a eukaryotic cell.

molecular assemblies became enclosed in such a vesicle to form primitive cell-like structures. The first cell membranes need not originally have been lecithin, for other molecules with hydrophilic and hydrophobic ends behave similarly.

There is, however, no evidence to prove that this is how it actually happened, and again we may never know, but it is a reasonable speculation. It is thought by many that, wherever there is water and conditions similar to those on Earth, then life may be found there.

In Box 8.2 the origin of mitochondria is discussed.

The situation then is that it is possible to speculate on how life may have arisen but we do not know how it actually took place. We do know that life is ultimately dependent on the properties of atoms which enabled them to assemble into living complexes. How they came to be as they are is at present an unanswered question.

Index